C0-AVY-247

Hello Tiny World

An enchanting journey into the world of creating terrariums

Hello Tiny World

Ben Newell

Discovering tiny worlds 06

Principles of terrariums 08

Terrariums to get started 50

Expanding horizons 84

Inside the bubble 118

All about plants 140

Cultivation and care 202

Resources 218

Index 220

Acknowledgments 223

About the author 224

Discovering tiny worlds

When I discovered terrariums, in 2016, I had no idea about the journey these tiny ecosystems would take me on. From being a teenager struggling with addiction issues, to becoming one of the world's most famous terrarium content creators, *Hello Tiny World* is the product of many years of repeated experimenting and failing.

My mother was born in Malaysia to Chinese and Taiwanese parents, but she grew up in Singapore, and it's here that my first memories of being around plants were born. Playing with my cousins in the Singapore Botanic Gardens, I was in awe at how completely different the plants and trees looked compared to those back at home in Worcester. The sounds of the insects made the whole place sing at night. Singapore is a botanical wonder and spending time there as a child was truly magical.

My teenage years and early twenties were plagued by addiction and mental health issues, which stemmed from the loss of my father when I was eight, then my brother when I was 13. It would be an overstatement to say that one thing saved me, but taking on a small plot of land and discovering horticulture at the age of 24 was a huge catalyst in helping me to turn my life around, and it was the true start of my journey with plants.

As I began my foray into horticulture, my mindset started to shift, and focusing my full attention on my vegetable patch was instrumental in helping me to leave behind the damaging life that I had been leading. It was probably an unorthodox entry, and having no prior experience with plants I didn't really know what I was doing. I've always learned better by working hands on, though, rather than in a classroom, and persisting with my vegetable patch was an important decision that changed the course of my life.

From here, all of my spare time was spent obsessing over plants; looking up unusual fruits and vegetables that I could grow on my vegetable patch, and reading about them in every spare moment I had.
I was discovering pockets of horticulture that I'd never been exposed to; topiary, cloud pruning, container gardening, growing herbs, houseplants, succulents, bonsai (which was of particular significance), and the protagonist of this story, the terrarium.

There have been eight years between getting my vegetable patch and writing this book, and a lot has happened in that time. My love for plants and terrariums has helped me through tough times and given me incredible experiences I never believed I'd have, such as winning a gold medal at the RHS Chelsea Flower Show, speaking to an audience of 1,000 people at the Bloomsbury Festival, and working with Disney and Marvel. Today, I'm delighted to have the honor of writing the book that I needed when I first discovered this hobby. For you.

Hello Tiny World will teach you the fundamentals of terrarium building so your journey (unlike mine) gets off to the strongest possible start. It contains an incredible selection of plants, and shows you how to use them for maximum effect. There are projects for you to undertake, too, and I encourage you to be creative, and add your own personal spin to them.

Whether you make only one terrarium, or they become an integral part of your life, I want this book to be the reference point you can turn to for the tools and steps you need to begin your own journey in this wonderfully rewarding hobby.

I hope they can do as much for you as they have for me.

1

Principles of terrariums

A permeable world

I often look through comments on social media posts to get ideas for new content that might be interesting or useful for me or my followers, and something that always seems to arise is the comparison of a terrarium to our own planet.

It may seem simplistic to draw comparisons between a human-made item like a terrarium and the intricate entity that is our Earth, but similarities undeniably exist. Exploring these connections brings a level of interest and creativity to the terrarium-gardening hobby, while also reminding us of the profound interconnectedness and responsibility that we have toward the wider natural world.

The main point of comparison is that both the Earth and a terrarium function as closed systems. A terrarium creates its own self-contained ecosystem that includes water, nutrients, and gases, which are all undergoing cycles and processes that create an environment suitable for sustaining plant and sometimes animal life. The Earth also operates in this way, whereby the amount of matter in it remains relatively constant, while allowing light to enter and leave the planet. It also experiences material cycles such as the water cycle, carbon cycle, and nutrient cycle, which are all critical for supporting life within its ecosystem.

Both on our planet and in a terrarium, different components (such as plants, animals, soil, light, and water) are all interconnected, relying on each other for survival, so a fine balance is required to create a stable habitat for the organisms within. It's important to recognize the vulnerability of the ecosystems of both a terrarium and the Earth. Just as external factors can easily disrupt a terrarium, the Earth's ecosystems are susceptible to disturbances caused by human activities—such as climate change, habitat destruction, and pollution—which pose significant threats to its well-being.

OPPOSITE Terrariums are an opportunity to create beautiful, tiny, self-contained ecosystems that reflect natural habitats from all around the globe.

What is a terrarium?

The word terrarium is used to describe a range of enclosures, but what specifically makes something a terrarium? Are the open-topped arrangements that have cacti and succulents in them considered terrariums? What about if it contains animals? The term bioactive terrarium has become popular recently, but what does it actually mean?

My personal opinion is that a container needs to have a high level of humidity to be considered a terrarium. Sometimes we use a lid to create this humid environment, or sometimes we can leave the top open if the opening is small and there is no detrimental impact on the level of humidity inside.

Naturally, I've explained many times at events, during workshops, and at talks, what I think a terrarium is, and rather than get into a long explanation about the multiple kinds and their differences, I prefer to simplify it.

A terrarium is a clear glass container that houses plants. Within the environment—whether it is sealed with a lid or left partially aerated with a small opening—a miniature water cycle occurs. As water evaporates from the substrate (the growing medium), and transpires from the leaves of the plants, it condenses within the container, creating a humid environment for tropical plants and mosses to grow in.

However, this explanation doesn't begin to convey the vast world that terrariums encompass.

LEFT Small-leaved plants such as *Saxifraga stolonifera* 'Mini' and aquatic moss such as *Taxiphyllum barbieri* (Java moss) are great choices for small terrariums like this candle holder that comes with a lid.

Types of terrariums

***terra*. From the Latin, meaning "earth"**

There are so many variations and permutations of terrariums that are constantly being blended, merged, and expanded on by creators and hobbyists. So let's go through some of the most popular types of terrariums and see what most appeals to you.

OPEN TERRARIUM

In an open terrarium, a transparent container is typically used to create a small-scale indoor garden using arid-region plants such as cacti and succulents. The container for this type of terrarium is not completely sealed, allowing for airflow and moisture evaporation.

Open terrariums are typically filled with a gritty growing medium that provides adequate drainage. Plants such as cacti, succulents, or other low-maintenance species are chosen based on their ability to thrive in drier conditions.

The key aspect of an open terrarium is the ability for moisture to escape and evaporate, preventing a build-up of excess humidity. This means open terrariums require more frequent watering compared to closed terrariums, which should be carefully administered (see page 15).

Aquariums

An aquarium (*aqua*, from the Latin, meaning "water") is a fully underwater system that contains aquatic plants and often fish and/or shrimp, too. I have experimented with keeping Walstad method plantings, which is where organic soil on the bottom of the container is held in place with a layer of gravel or sand. No technology is used to filter the water in the container—the plants keep it clean. If I had more space I would love to create more aquariums.

For practical reasons, and due to a lack of expert knowledge about the subject, I won't be talking about aquariums in this book.

FULLY SEALED TERRARIUM

Many of the plants that we use in our terrariums come from parts of the world that experience 80 percent humidity and higher. However, they also grow outside where they have access to infinite amounts of air. A terrarium provides sufficient humidity, but it can become stagnant and lack any kind of airflow. For this reason, I don't like to fully seal terrariums, as airflow is a key component to terrarium health, so I only create terrariums like this as an experiment. The terrarium created by David Latimer in 1960 (see box below) is a good example of how a sealed terrarium can work, but it's worth noting that his project is an anomaly; rarely do terrariums survive anywhere near this length of time, and I put its success down to the size of the container and the amount of light it's receiving. Have fun experimenting with these, but go in with a low expectation of success.

OPPOSITE Cookie jars can be used as fully sealed terrariums; however, for the best chance of success I recommend removing the lid from time to time to increase airflow and to do general maintenance.

MOSSARIUM

A mossarium is a specialized container that is designed to create a suitable environment for growing and displaying mosses. It is similar to a terrarium or vivarium (see page 17), but it differs in that it is specifically focused on providing optimal conditions for moss growth.

I recommend using aquatic species of moss over terrestrial in mossariums because there is a broader range suitable for this specialized environment. These are very easily purchased because they are more widely available from aquarium plant specialists and are the most sustainable options.

David Latimer's terrarium

In 1960, electrical engineer David Latimer created a terrarium that I'm sure he had no idea would go on to become one of the most famous terrariums in the world. The container is an old wine demijohn, which is very large, inside which he placed some compost (not specific terrarium soil), water, and a single *Tradescantia* seedling. For the next 12 years, the *Tradescantia* grew happily in the container and it was in 1972 that David opened the container for the first, and last, time to give it a drink of water.

Since then the demijohn has remained shut and has become a rare case of a true self-sustaining ecosystem. According to David, it sits 6ft (1.8m) away from a bright window and the plants grow toward the light (phototropism), and all he has to do is rotate the terrarium to ensure even growth.

RIPARIUM

ripa. From the Latin, meaning "bank" or "shore"

A riparium is a type of aquatic or semiaquatic setup that is designed to mimic the transition zone between land and water, which is typically found along the banks of rivers, streams, or wetlands. It is a specialized aquarium or enclosure that combines elements of both terrestrial and aquatic environments, creating a unique and visually appealing display.

In a riparium, a portion of the enclosure is dedicated to water while the rest is reserved for emergent or marginal plants. The water section can contain aquatic plants, fish, and other aquatic creatures, while the land or emergent section hosts plants that can tolerate wet conditions or submerged roots.

The setup typically includes a tank or container with a land section that is partially or fully submerged. The water is often filtered, and appropriate water parameters, such as temperature and pH, must be maintained to support any aquatic life. The land section, usually located at the edges of the container or above the waterline, is typically filled with an aquarium substrate suitable for plant growth.

Ripariums often feature a variety of plant species, including both fully aquatic plants that grow entirely submerged in the water and emergent plants that can thrive with their roots partially or periodically submerged. These setups can also be populated with a range of aquatic organisms, such as small fish, shrimp, or snails, depending on the size and design of the enclosure. The plants are often rooted underwater but their leaves can emerge above the water line. They can also grow on emergent branches or rocks.

PALUDARIUM

palus. From the Latin, meaning "marsh"

A paludarium is a vivarium or terrarium that merges both terrestrial and aquatic habitats. Unlike a riparium, which highlights the boundary where water meets land, a paludarium has clear terrestrial and aquatic zones. In a riparium, the focus is on plants that grow semisubmerged or just above the waterline. A paludarium can also have this feature but it has a significant proportion of the tank dedicated to terrestrial plants.

In a paludarium, a portion of the enclosure is dedicated to water, while the remaining area is set up as a terrestrial environment. The water section can range from a small pool or pond to a larger body of water, depending on the size of the container. The land section typically features a substrate that is suitable for plant growth, and can also include a variety of plants that tolerate damp or periodically submerged conditions.

The water in a paludarium is often filtered and maintained to support aquatic life. It may contain fish, amphibians, invertebrates, or other aquatic organisms that are well suited to the environment. The land section is designed to provide a suitable habitat for various terrestrial plants and occasionally animals such as reptiles, amphibians, or invertebrates that can thrive in both land and water environments.

Paludariums often incorporate features such as rocks, driftwood, branches, and other natural elements to create a visually appealing and realistic representation of a wetland or swamp ecosystem. These features provide climbing surfaces, hiding places, and resting spots for the inhabitants of the paludarium.

The key aspect of a paludarium is the careful balance of both terrestrial and aquatic elements. It allows for the creation of a diverse and dynamic ecosystem, showcasing the interactions between plants, animals, and the surrounding environment.

VIVARIUM

vivus. From the Latin, meaning "alive"

A vivarium is usually a planted terrarium that contains an animal such as a dart frog, gecko, or spider. They are designed to mimic the natural habitat conditions of the animals they house. For example, a dart-frog enclosure will be planted to simulate the rainforest floor; a crested-gecko enclosure will be designed to replicate an environment in the trees (as these creatures are arboreal in nature); and a tortoise vivarium will simulate a desert environment with little to no vegetation. A terrarium doesn't need to have live plants inside to be considered a vivarium, but using artificial plants severely impacts the overall aesthetic.

Proper research and consideration of the specific needs of the organisms are essential when setting up and maintaining a vivarium, to ensure the well-being and health of the inhabitants. In particular, it's important to note that each vivarium must be specifically tailored for the animal it houses, as each animal will require different levels of humidity, ventilation, and lighting, and have varying feeding regimes.

BIOACTIVE TERRARIUM

bio. From the Greek *bios*, meaning "life"

These are the types of terrariums for which I've become known (see page 18)! Bioactive terrariums are closed containers with ventilation that host a cleanup crew in the form of isopods and springtails, although millipedes and snails can also be used. The function of these custodians is to break down any waste matter, which in turn fertilizes the plants. The custodians also aerate the substrate, which is beneficial for the plants.

A bioactive terrarium is a true ecosystem because the plants and custodians have a symbiotic relationship; the healthy growing plants produce oxygen, which is important for the creatures, and they also shed older leaves and roots, which feeds the animals. As the animals eat, they defecate into the soil, which breaks down and becomes available as nutrition for the plants, helping them to grow healthy and strong.

RIGHT *Armadillidium* isopods are a great species to get started with when making bioactive terrariums.

The most famous terrarium in the world?

In August 2021, I made a terrarium in a 16in (40cm) bowl using a large piece of cork bark as a centerpiece, surrounded by *Ficus colombia* and *Ficus punctata* cuttings, a *Biophytum sensitivum*, a *Nephrolepis cordifolia* 'duffii' fern, and an *Asparagus setaceus* (asparagus fern). I added a colony of *Cubaris murina* woodlice, some *Desmoxytes planata* (pink dragon millipedes), and *Folsomia candida* (springtails). The creatures have a close relationship to the plants; the bigger and healthier the plants, the more oxygen they produce; the more oxygen that is produced, the happier the creatures. The more they feed, the more they poop, and the healthier the soil becomes.

The animals wouldn't survive in a regular terrarium, so the environment must be carefully tailored to their needs. A large amount of leaf litter replicates the forest floor and is topped up regularly to act as a long-term food source; woodlice and millipedes nibble away at decomposing litter, and as it decomposes further, the springtails start to feed. The pink dragon millipedes need ample rotting wood—their main food source. It's also helpful to provide fresh vegetables and fruit, which the creatures will quickly devour. Millipedes and woodlice need supplementary calcium and protein to keep their exoskeleton strong and healthy, which can be provided by a light dusting of cuttlefish bone and fish food.

An unintended addition here was the tiny snail *Oxychilus alliarius*, or garlic snail, when a few eggs accidentally found their way in. Within a few months there were vast amounts of them. Usually considered a pest that nibbles on plant leaves, the snails didn't have much to feed on here, because *Ficus* plants exude an irritable, milky latex when cut, so the snails avoid them. The extremely fine leaves of the asparagus fern also don't appeal to the snails, nor the tough foliage of the *Nephrolepis* fern.

As time has gone on, this terrarium has developed and now hosts a healthy colony of creatures. The millipede population dropped off but it has bounced back; the woodlice, while evasive, are holding steady numbers; and the springtails are plodding along happily, doing their bit. While it is an ecosystem, the terrarium performs far better with assistance. This is why I steer clear of calling any terrarium a self-sustaining ecosystem, as what happens when the plants get too big? Or there isn't enough food for the creatures? Or if the soil becomes dry or there is a mold outbreak?

Calling this the most famous terrarium in the world is some claim, but the videos I made that recorded the creatures feeding have collectively gained hundreds of millions of views across my social media platforms. When I took this terrarium to the RHS Chelsea Flower Show in 2022, countless people recognized it. I've even been stopped in the street by people telling me they've seen the videos! There is no quantifiable way to claim this is the most famous terrarium—I hope it's up there, along with David Latimer's (see page 15).

Soil

Those of us who have grown plants over the years will all agree that there is something quite amazing about soil. I remember adding countless wheelbarrows of well-rotted horse manure to my old vegetable patch, which resulted in an incredibly rich, well-draining, airy soil that was teeming with life. Just before I finally left the site, I came back to remove the mulberry tree (*Morus nigra*) and the Sichuan pepper tree (*Zanthoxylum simulans*) I had planted there. When I dug them out of the ground, I found that the soil was dry, hard, and lifeless.

This showed me firsthand the need to feed your soil regularly—whether with compost, manure, or leaf mold—as plants can consume nutrients fast. I have also learned that the same principles that apply to my vegetable patch also relate to terrariums. As closed ecosystems, it's so important to use a mix that suits the unique terrarium environment, so that this growing medium enables healthy growth and supports plants over a long period of time.

In an ideal world we would tailor the soil, or substrate, mixes to specific plants, but because we usually mix plant species in a terrarium, it's better to use a general terrarium soil mix. There are many options, and what you choose is often dependent on personal preference and availability, but I must stress the importance of using a high-quality substrate. Skimping on this will mean problems later.

What makes a good-quality terrarium substrate?

You'll hear the term "water-retentive yet well-draining" used often to describe terrarium substrate, but what does it mean and why does it create ideal growing

RIGHT If you are making just one terrarium, buying a premix substrate is fine, but as you get further into the hobby you may prefer to tailor your own mixes to your terrarium projects. The simple mix (left) is a basic substrate, while that on the right is a more premium option (see page 31).

conditions? Essentially, water-retentive means the medium can hold on to water for periods of time, which increases its availability to the plants as and when they need it; whereas well-draining soil allows water to percolate through it, preventing standing water rotting plant roots. You really want a little of both—to ensure excess water drains away but the soil holds on to enough water to prevent the plants experiencing and suffering from drought conditions. In a terrarium we have to be careful when watering to add just the right amount, due to the lack of drainage holes in the container.

To achieve this balance, I find altering the ratios of components within the soil and experimenting with what's available really useful (see pages 24–30), to help boost nutrients or improve the substrate's water-retaining capacity. For example, compost and coir on their own are far too water-retentive, and within the confines of a terrarium they can become hydrophobic (they stop taking in water) and potentially anaerobic (lacking oxygen), which prevents the roots "breathing" and often ends in their demise.

While bonsai soil mediums are extremely water-retentive, on their own they are often too airy and well draining. To remedy this, I combine compost and/or coir with a few optional extras to create a free-draining but water-retentive substrate. The plants' roots will not grow directly into the bonsai mediums, rather, they'll grow on them and in the spaces between the grains. Filling these gaps with compost, coir, and worm castings ensures the roots will grow healthily.

Which substrates to avoid

Before we get started on which are the best soil mediums or combinations for terrariums, it's worth discussing those that should be avoided.

GARDEN SOIL

Everybody's garden soil is different, so I'm reluctant to use a blanket term here, but for the best chances of success I recommend steering clear of digging it up to use in a terrarium. As there are so many better alternatives, I see no reason why you'd need to use this soil; it hardens quickly inside a container because it loses nutrition and becomes hydrophobic.

In some experiments I have made terrariums using collected soil, but if I do use this, I always take it from an area that has had heavy leaf fall—and even then I only remove the top few inches. This means the soil is basically leaf litter combined with rotting twigs and wood, with minimal amounts of topsoil. This can work well, but I would still recommend using a homemade or high-quality premix instead.

PEAT

Due to the unsustainable methods of harvesting peat, I don't use it at all and I highly recommend you avoid soil mixes that contain it—check for its inclusion on all growing media packaging.

COIR

This material, the outer husk of a coconut, is becoming increasingly popular as a peat alternative. While I like to use coir as an addition to a terrarium substrate, on its own it's a poor choice for numerous reasons, mostly because it's devoid of nutrients. It also becomes hydrophobic quickly if it is allowed to dry out, which means it then needs to be re-saturated.

CHEAP PRE-BOUGHT MIXES

The vast majority of substrates sold in garden centers that are labeled "terrarium soil" are just bags of compost with a little perlite added. If you do buy these, I recommend adding a mix made from 1 part shop-bought to 2 parts of the mediums listed on pages 24–30.

Sterilizing substrates

While some guides suggest you sterilize natural hardscape materials for your terrarium (see page 93), this is something I'm strongly against for substrates. Having a healthy network of mycelium and microfauna within the substrate is beneficial—we should be encouraging the life in our soils, not killing it! So, please don't sterilize growing mediums.

Which substrates to use

Before delving into the best soil, here's a list of high-quality mediums you can use to help create healthy terrarium conditions. These are the base substrates, to which you can add certain components to boost nutrients or water-retaining capacity (see pages 24–30).

COMPOST

When combined with other components, a good-quality compost does a great job of providing nutrition and microbial activity. If you have a well-aged compost heap in your garden, it'll be teeming with beneficial life! Take a handful and observe closely, the compost should be dark in color and will feel airy and springy, with a lot of body to it. You may even be able to spot decaying bits of organic material, or some composting worms and worm eggs!

True, high-quality compost is difficult to find, and the vast majority is sterilized upon packaging, stripping it of the beneficial life that makes it so good. I recommend steering clear of cheap options, as good compost is a labor of time and love, and so often comes at a high price. For those in the UK, Compost Club sells amazing compost that works especially well in terrariums.

PREMIXED TERRARIUM SOILS

When buying a premixed substrate, choose one from a reputable company. I've found two suppliers in the UK who sell top-quality terrarium substrates: Grow Tropicals and Soil Ninja (who also supply throughout Europe).

Grow Tropicals have two options: Simply Terrarium and a terrarium mix based on the Atlanta Botanical Gardens (ABG) mix—see right. Soil Ninja have a general terrarium substrate that is great for use in terrariums of all sizes, however, I prefer to use it in larger builds, due to its water-retentiveness.

If you are struggling to find some in your area, look for a high-quality desert and cacti mix. These vary widely both in quality and composition, but in many cases just adding a few parts of compost, worm castings, or coir to the mix helps improve its water-retentiveness.

ABG MIX

This is a specialist mix originally created in the Atlanta Botanical Gardens, in the US, which has become the gold standard of substrates used in closed containers. It's well-draining, but has great water retention, is airy, and doesn't compact. You can be sure that any medium combination based on this mix will be of good quality; however, do check the packaging information, because the original mix uses peat moss. Newer alternatives that are now available use coir or composted bracken as a peat alternative.

SPHAGNUM MOSS

Although devoid of nutrients, which means it's not ideal to use on its own for a prolonged period, sphagnum moss is great for propagating. It is available alive or dried; both are suitable for terrariums, but when using dried you'll need to rehydrate it before use. There are different grades of sphagnum moss and you get what you pay for —cheaper moss needs sifting to separate out leaf and twig debris before you can propagate with it, whereas premium options are debris-free and ready to work with.

I recommend chopping (or "milling") the sphagnum into smaller pieces before using it either on its own or as a component, as I prefer the way it sits in the substrate when it's milled. You can do this with your hands, scissors, or a knife.

I have recently discovered Dusk Moss Mix, which is milled sphagnum moss with added fern and tropical moss spores. It is sold in packets and can be applied to backgrounds and on top of substrates, where, in time, it grows into a lush green carpet of moss that has random ferns and occasionally other tropical plants germinating from it. Keep it damp and under good lighting (minimum 200fc) for optimum growth.

Recommended components

The best terrarium substrates are composed of multiple materials that can be combined to create a growing medium that is free-draining yet water-retentive, airy and not compacted, and nutritious. I particularly like using bonsai substrates, which are of a really high quality. I recommend using the finer-grade mediums, where the grains are between 1 and 5mm, because they are more water-retentive, but you can add in larger grains, too, which will improve the drainage. Any gaps between the grains can be filled with coir, compost, or sphagnum moss—or all three. Mixing and matching is the key to creating the optimum substrate, so here are some of my favorite components. All the materials listed here are options for successful growing media; some are more ecologically favorable than others—some have a high carbon footprint, others may require intensive production. Whenever you buy substrates, do check their sustainability.

AKADAMA

The most well known of the bonsai mediums, akadama is highly regarded for its unique properties that make it ideal for use in bonsai; it has excellent water-retentive qualities, while also providing good drainage. The granular structure of akadama creates an aerobic, well-draining environment for the roots, which promotes healthy development and prevents root rot.

Akadama is often used as a component in soil mixes for bonsai trees, succulents, and other container plants. It's typically combined with ingredients such as pumice, lava rock, moler clay, and organic matter to create a soil mixture that's tailored to the specific needs of the plants being grown. It has a neutral to slightly acidic pH. While akadama is widely used in Japan and many parts of the world, its availability can vary depending on your location. It's becoming increasingly expensive to source, but thankfully there are other alternatives available, such as the ones listed in this section.

Sakadama is a bonsai medium made from kiln-fired clay. It's fired at 1112°F (600°C) to create a hard grain that lasts a long time in all weather conditions. Thankfully, in a terrarium the substrates don't freeze, which means this will last indefinitely! It's lightweight and has good water-retentive and draining properties. It's a good substitute for akadama.

MOLER CLAY

Moler clay (or molar clay/calcined clay) is such a versatile material, and its popularity in horticulture is growing. It is a type of diatomaceous earth found in moler deposits in Denmark and Sweden, and is lightweight due to its porous structure. Moler clay has a high absorption capacity and can retain water and other liquids really well, making it useful in industries like agriculture, horticulture, and water treatment. It's used by mechanics to clear up diesel spills and by pet owners as a cat litter, too! I think it makes a wonderful practical addition to terrarium substrate.

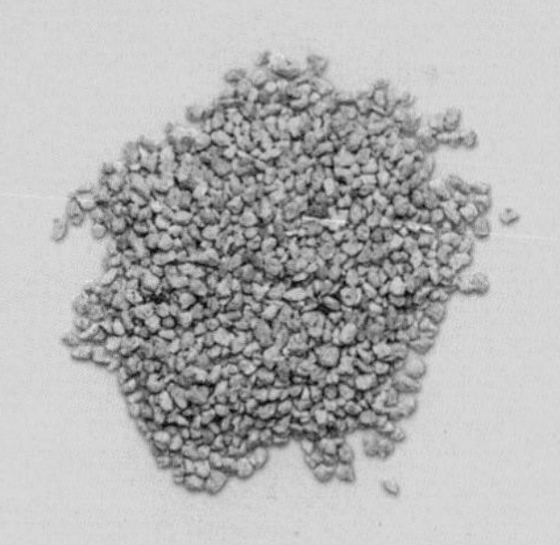

PUMICE

A type of igneous volcanic rock, pumice is one of the lightest rocks found in nature. It forms when lava solidifies quickly and traps gas bubbles within its structure. Its porous structure is due to the high volume of these gas bubbles, which makes it lightweight. It provides excellent drainage and soil aeration, preventing soil compaction and promoting healthy root growth. Pumice has a neutral pH. It's a near-perfect medium but its white color means it stands out in arrangements. Fine pumice is available from all good bonsai suppliers.

LAVA ROCK

My favourite bonsai medium is lava rock, sometimes referred to as lapillo, which is a volcanic rock. During volcanic eruptions, particles explode into the atmosphere and rapidly cool on their descent. The solidified volcanic lava stones that are produced have a porous honeycomb structure, making them extremely water-retentive and well draining. Lava rock has a neutral pH, which means it will not affect the acidity or alkalinity of your soil, and has a pleasing color.

Lava rock can be mixed into soil, but it is also useful as a drainage layer at the bottom of a terrarium or as a top-dressing—aquatic mosses grow especially well on it.

KYODAMA

This fired-clay, human-made product is excellent for providing drainage in soil mixes. It has a neutral pH and a very attractive light color. Due to its gritty nature, kyodama is often used in bonsai but also commonly used in cacti and succulent plantings. While it has good water-retentive qualities, I find it works better at improving drainage. It's a hard grain that doesn't break down over time, meaning it can be sifted out and used again when you are replacing your growing medium.

ACTIVATED CHARCOAL

Made from carbon-rich materials, activated charcoal, as its name suggests, undergoes a process called activation. This is when it's heated at high temperatures and treated with steam or chemicals to create a highly porous structure.

The activation process increases the surface area of the charcoal pieces and enhances its absorption qualities. This material is granular in form and can be added directly to the substrate or used in a layer. Because of its high adsorption capacity (the ability to hold onto gas or liquid molecules as a thin film), it effectively traps and binds toxins, chemicals, odors, and impurities, making it a useful addition in a closed terrarium.

Activated charcoal does have a high price point, and the fact that practically all of my terrariums don't contain it and still perform exceedingly well means I see it as a beneficial but not entirely necessary component. A more cost-effective option would be to use organic barbecue lumpwood charcoal. This comes in larger pieces that need to be broken down a bit, and because it's very dusty when split, I advise soaking it in water for ten minutes, then placing it in a plastic bag and using a hammer to break it into smaller lumps. It will still be pretty dusty work, though, so do this outside!

It's important that any charcoal you use is free from catalysts; if it has a catalyst the packaging will brand it "easy-light charcoal."

COIR

When the fibers of the outer husk of the coconut are ground down we get coir (also known as coconut coir/coco coir). It's a product that's used as an alternative to peat and is often referred to as being environmentally friendly. However, questions have recently arisen about the methods of harvesting and how sustainable it really is, due to the contribution of palm-tree plantations to deforestation, the high water consumption used in the processing methods, and the transportation issues derived from it coming from tropical countries. Coir can be sourced from suppliers who employ sustainable farming practices, though, and from those who work to reduce carbon footprints in transit.

I wrote about coir earlier and how it's a bad medium to use on its own, as it can become hydrophobic (see pages 21 and 22), but as a soil addition it really shines. The way the bonsai mediums naturally sit against each other creates air pockets between the grains that can hasten the drying of the substrate, so filling that space with coir improves water retentivity and the overall body of the substrate.

COMPOST

This nutrient-rich material is created through the decomposition of organic matter. It's commonly used as a soil amendment and a fertilizer in gardening rather than a growing medium alone.

In a terrarium environment, compost should be used in the same way as coir; as an addition to other mediums. Homemade, high-quality compost contains masses of beneficial microfauna that contribute to the terrarium ecosystem. Unlike coir, compost contains a wide range of essential nutrients that are released slowly and are made available to plants over time. It helps improve soil fertility and provides a healthy nutrient supply. On the opposite end of the scale, most cheap supermarket-bought compost has little to no nutrition and can be harmful to plants.

I advise thoroughly washing your hands after handling compost—just to be safe—or you can wear gloves, if you prefer.

EXPANDED CLAY PEBBLES

Expanded clay pebbles, also known as supalite, hydroton, or LECA (Lightweight Expanded Clay Aggregate), are light, porous clay balls that are used as a growing medium in hydroponics and aquaponics and as a soil addition or drainage layer. They are created by heating clay pellets in a rotary kiln, which causes them to expand and form a honeycomblike structure.

These expanded clay pebbles are lightweight due to their porous structure, which makes them particularly useful in a terrarium, especially when used as a drainage layer at the bottom. Their porosity also makes them a great addition to a terrarium substrate, as it allows for good aeration and drainage, promoting oxygen availability and preventing waterlogging in the root zone. These pebbles are typically pH neutral.

Fertilizer

It's often stated that terrarium substrates need to be fertilizer-free to stop the plants growing too quickly, but I disagree. Healthy plants are naturally more resistant to pests and disease, and part of growing healthy plants is to provide them with a medium that is rich in nutrition. I'd much rather have healthy plants that are growing fast than slower-growing, malnourished plants. You just need to prune more often.

HORTICULTURAL CHARCOAL

Typically made from hardwoods such as oak or maple, horticultural charcoal is produced through a process called pyrolysis, which results in charred wood that is rich in carbon composition. Horticultural charcoal is not activated (see page 26) and has a lower porosity compared to activated charcoal, and it helps improve soil drainage by creating air pockets in the substrate, which allows excess water to drain more efficiently and prevents waterlogging.

Although less porous than activated charcoal, horticultural charcoal can still absorb excess moisture and impurities that are in the substrate. However, its primary function is around soil improvement and plant health rather than having strong adsorption qualities.

The function of charcoal in a terrarium is often debated. I'm unsure of how effective it is at removing toxins and bacteria from the soil, but I do like the way it can improve soil structure.

LEAF MOLD

I love leaf mold! The 12-acre garden I once worked on was surrounded by deciduous trees, so there was always a pile of leaf mold made from fallen leaves that had been collected years before. It was the most humus-rich, airy medium that almost crunched in your hands; it was full of life and I took great pleasure in digging it into the vegetable beds to improve the soil structure.

Leaf mold is formed by microorganisms, fungi, and worms breaking down fallen leaves. This decomposition process can take several months to two years—or longer—depending on the temperature, moisture levels, and types of leaves. The finished leaf mold is dark brown or black in color and has a crumbly texture that looks nothing like the leaves it came from. It is rich in beneficial microorganisms but doesn't contain much nutritional value, so it is used mainly as a soil amendment or mulch.

I love using leaf mold in bioactive terrariums and I've noticed that the custodians spend a lot of their time in it!

TREE FERN FIBER

This material comes from the trunk of certain species of tree ferns, such as *Dicksonia antarctica*, which are large, tropical ferns that grow in forested areas.

The trunk of a tree fern is made from densely packed layers of old leaf stalks and organic matter. This fibrous material is highly porous and provides excellent aeration and drainage when mixed into a terrarium substrate. Its ability to withstand high humidity and retain moisture makes this a perfect medium for a mounting material for epiphytic plants, while the fibers also provide a supportive, moisture-retentive substrate for plants' roots to anchor into.

There is some controversy about using tree fern fiber as a growing medium because of sustainability concerns, in particular over how it is harvested. However, there are suppliers who only source from privately owned forests, which is a better, more sustainable option. I've never bought tree fern fiber, but I was lucky enough to have some freshly harvested fibers dropped off to me by a supportive social media follower! It's a great material that has multiple uses in a terrarium setting, but please do research its origins before buying it.

WORM CASTINGS

Sometimes called vermicast, worm castings are the droppings produced by composting worms as they digest organic materials. It is a fast way of producing a high-quality compost and is a type of natural fertilizer that is highly beneficial for plant growth and soil health. I like to use a small amount mixed in with substrate, then a few months later I add a little more as a top-dressing.

It's important to buy pure worm castings that are not mixed with anything else. Fishing tackle shops often breed these worms for bait, producing a byproduct of worm castings. If you're happy to have a few thousand worms living in a container in your garden, a wormery is a great way of turning your waste into free, nutrient-rich castings for your garden and terrariums.

ZEOLITE

This material is unique in that it has cation exchange properties, meaning it can attract and retain certain nutrients within its structure. It holds on to these essential nutrients and makes them available to the plant as the roots come into contact with the zeolite particles. Zeolite absorbs excess nutrients and reduces the risk of chemical burns on plants that are artificially fed.

Like the other bonsai materials, zeolite has good water-retentive and draining qualities, and I like the way it looks in the substrate—it starts off white but changes to a dull beige when wet.

ORCHID BARK

This typically consists of small pieces of tree bark, and adds structure to substrates, provides drainage, and has excellent water-retentive properties. As the bark breaks down it adds nutrition into the substrate. Orchids have unique root structures that require a well-draining growing medium that allows for air circulation around the roots. Orchid bark provides these conditions by creating open spaces between the bark chunks, which also makes it a great addition to a terrarium substrate. Orchid bark is sourced from coniferous tree species such as Douglas fir (*Pseudotsuga menziesii*), pines (*Pinus* sp.), or cedars, or other hardwood trees such as oak (*Quercus*), maple (*Acer*), or beech (*Fagus*).

Sand

I rarely use sand in my substrate mixes as I don't think it's necessary, but I have noticed it is present in some of the shop-bought mixes that I've tried. I do like the way that the tiny grains complement the other mediums and how it helps to improve overall structure and drainage. It's important to only use it in small quantities, though, as the tiny grains can clump together when wet and hold on to too much water, which has an adverse effect on the drainage qualities of the medium.

How to make your own terrarium substrate

Here are three mixes that I frequently use. Don't worry if you're missing a few items for these; the recipes below are just guides and you can add any materials that are available to you. I tend to use the simple terrarium mix in smaller terrariums and the premium in larger ones.

I prefer to make up this substrate as needed, because I like to keep the worm castings and compost moist, and the other materials dry. You can make a big batch of this if you like, just make sure it doesn't fully dry out, as this will negatively affect the microbial activity in the worm castings or compost.

Simple substrate

Easy to prepare, this mix is suitable for terrariums of all sizes.

MATERIALS

2 parts bonsai medium (such as moler clay, akadama, lava rock, pumice)
2 parts compost or coir (or a mix of one each)
1 part worm castings

Mix all the above mediums together thoroughly.

Substrate for large terrariums

This is a specialized mix suitable for large terrariums or vivariums. I've based this on the ABG mix, but I've swapped the peat for coir and added worm castings for nutrition.

MATERIALS

2 parts tree fern fiber
2 parts sphagnum moss
1 part worm castings
1 part coir
1 part orchid bark
1 part activated charcoal

Mix everything together thoroughly.

Premium substrate

This premium terrarium mix is suitable for terrariums of all sizes.

MATERIALS

1 part sphagnum moss
1 part akadama
1 part lava rock
1 part moler clay
1 part composted pine bark
1 part tree fern fiber
2 parts high-quality compost
1 part worm castings

Chop or tear the sphagnum moss into small pieces, then mix them thoroughly with the other materials in a well-ventilated area.

NOTE If using coir instead of compost, use 2 parts worm castings.

Water

Water is a vitally important component in the growth of plants, and getting the level right in a terrarium is crucial for success—dry soil leads to wilting plants that flop and die unless they are swiftly watered, while wet soil quickly becomes anaerobic (lacking oxygen) in the sealed environment, which will lead to the plants' demise.

There is a point during my in-person workshops where I try to convey how to water a terrarium, which often results in confused faces. People want a simple numerical answer to this question, and they can be disappointed when they don't get one! The reason for this ambiguity is that there are so many variables within the ecosystem of a terrarium that determine how much water needs to be added and how often.

When does my terrarium need watering?

If there is one piece of advice I would give, it's do not water to a schedule. What I mean by this is setting dates on your calendar to water (such as once a week, once a fortnight) is a bad idea. A terrarium needs watering when it needs watering, and unlike houseplants, there is no forgiveness if you overwater a terrarium, because there are no drainage holes for excess water, so it will just collect and pool in the bottom.

OPPOSITE I tend to mist this terrarium containing *Pyrrhobryum dozyanum* moss, *Mnium hornum* moss, and *Ficus thunbergii* around once a month.

The key is to remember that when it comes to watering a terrarium, less is more. But if you're not sure, a good way to tell when you need to add more water is through observation of the substrate or by physically touching it.

If you've overwatered your terrarium, first try to remove as much of the sitting water as possible. You can do this by carefully tipping it out, but make sure you hold the plants in place as you tilt! Then, using a piece of paper towel or a microfiber cloth, try to absorb some of the excess water. This can work in minor cases, but in more major overwatering issues, using a turkey baster to siphon out the excess water can be more effective.

For terrariums that are severely overwatered, it's often best to start again. There's no turning back once a substrate is saturated, as the water won't evaporate quickly enough before it becomes anaerobic. This can be disheartening but view it as a learning curve. I overwatered so many terrariums before I learnt that "sweet spot," so don't give up!

Knowing when to water

Your terrarium needs water if:

- the substrate is light in color
- the plants are wilting
- moss is looking dry
- the terrarium feels light
- the soil surface is loose and moves easily
- the substrate is dry to the touch

Your terrarium is sufficiently watered if:

- the substrate is damp to the touch
- the substrate is darker in color
- the plants look healthy and are thriving

Your terrarium is overwatered if:

- the soil feels saturated
- water has pooled in the bottom
- there is an unpleasant smell
- you can see water move as you tilt the terrarium

Types of water

Tap water is the most obvious water to reach for, but it does contain harsh chemicals and leaves a calcium buildup on the glass of your terrarium that can be difficult to clean away. Avoid using this unless you have exceptionally soft water in your area.

If you have a water softener fitted to your faucets, please note that this water is not suitable for use in a terrarium, as the sodium from the filter will build up and can negatively affect plant health. These are my suggestions for good alternatives to tap water in order of preference.

DISTILLED

This is simply water that has been heated to create steam, which then condenses into liquid that is collected in another container to create a purified water. During this process any mineral or chemical components are removed. Distilled water does not leave any stains on the glass or mineral build-up in the soil. It is widely available and perfect for use in misting systems as it will not affect mechanical performance over time.

REVERSE OSMOSIS

In reverse osmosis, water is forced through a semipermeable membrane so that any contaminants become trapped in the filter. The result is a pure form of water that's perfect for drinking or for use in any horticultural setting. It is available from aquatic stores and will leave no mineral buildup on the glass or in the soil.

DEIONIZED

This is a cheaper, more easily available alternative to distilled water that uses ion exchange to remove almost all of the minerals from the water. I get my deionized water from my local supermarket; it's usually around $15 for a gallon, which will last for a few months.

FILTERED

Water filters do not remove all calcium but they do strip out the minerals that result in limescale. I often use filtered water in my terrariums, as it's convenient and a vast improvement on tap water.

BOTTLED

Many kinds of bottled water are of high quality, but there is still a mineral composition within it, meaning it can leave stains on the glass. Bottled water is an improvement on tap water, but it's still better to use the other alternatives listed here, if possible.

RIGHT Choosing the best-quality water you can access is key to healthy plants, and it is best sprayed onto the terrarium contents with a pressurized spray bottle, which offers speedy, accurate watering.

How to water a terrarium

Terrariums are closed ecosystems, which means they recycle much of the water inside them and only need the occasional additional watering. In fact, one of the most appealing aspects of keeping terrariums is precisely how little they need to be watered once created! Of course, container size, light level, temperature, the substrate used, and how established the plants are all play a part in how often you need to water, but compared to houseplants this is a far less frequent task. I've had some containers unopened and not watered in over a year!

SPRAY BOTTLES

Using a spray bottle to water enables a slower, more accurate, and careful watering of your substrate. I prefer to use a pressurized spray bottle to water all my terrariums; they're more expensive than standard spray bottles but they're faster and easier to use, saving a lot of time as they spray continuously. While normal spray bottles work well, in larger terrariums you'll find you are pressing that trigger a lot!

WATERING SMALL TERRARIUMS

In smaller terrariums, the mist setting on your spray bottle is ideal; position the nozzle into the opening and lightly spray until the substrate is three-quarters moist. Don't worry if a little seeps into the drainage layer (if you've used one), this is normal, but avoid creating any substantial puddles, as this is a sign that you've overwatered (see page 34).

WATERING LARGE TERRARIUMS

Larger terrariums will take forever to water on the mist setting, so use a setting between mist and jet. Do this carefully and ensure the substrate is evenly moist. Aiming the jet at the sides of the container will enable you to see how much of the substrate has received water. Once the top half is moist, stop and wait for five minutes. Let gravity pull the rest of the water downward, then add more if needed. It sounds obvious, but it is important to ensure the entire substrate area is watered, not just one part.

TOPPING UP

High-quality terrarium substrates containing bonsai mediums are extremely water-retentive, so after the first watering it's often only the surface that is sufficiently watered. Test the soil after the first watering before you start to build the planting to see how much of it is moist, then water again accordingly.

Once watered, your substrate should feel like a sponge wrung out of all its water. Imagine you're about to wipe down a dusty surface; you'd plunge the sponge into a bowl of water, remove it, then squeeze out the excess. The level of moisture left in the sponge is what we aim for in our terrarium substrate: slightly moist.

OPPOSITE Note how the darker area indicates the soil is damp, and the lighter area is dry. While the side view indicates how far the water has traveled downward, you must ensure the center of the substrate is sufficiently watered, too.

Misting moss

From terrestrial mosses like *Leucobryum glaucum* to aquatic species like *Vesicularia dubyana* (Christmas moss), which are capable of growing out of water, mosses are such a wonderful aspect of terrarium building. If you do decide to include them, there are a few things to keep in mind.

Unlike plants, mosses take the majority of the moisture they need from above soil level and do not have a root system. This makes a terrarium the perfect environment for many species, but you will need to keep a close eye on them. Sometimes the ambient humidity is sufficient to keep them growing healthily, although this isn't always the case, and observation is key to determining exactly when they need watering.

In larger containers with a lot of space above soil level, mosses will dry out more quickly, so they need frequent watering. In smaller containers, mosses will recycle water quicker, needing less frequent watering.

Bulkier mosses, such as *Leucobryum glaucum* and *Dicranum scoparium,* hold on to large amounts of water. You can prepare them for planting by fully submerging them in water, squeezing out the bubbles, then removing the moss from the water and squeezing out any excess water. You'll be left with a moist cushion of moss that's ready for planting.

Once the moss is in the terrarium and (hopefully) happily growing, you should water it carefully but substantially by misting the surface. The aim is not to wet the substrate around the moss but to dampen the surface. Be careful not to saturate your substrate when doing this. Finer moss species, such as Christmas moss or Java moss, have more delicate leaves and only need a light spray to be sufficiently watered.

Condensation

A common worry among new terrarium hobbyists is about the amount of condensation that appears on the inside of the glass of their containers. I've heard many people say this is due to an excess of moisture, but in my experience it's usually down to the terrarium getting a lot of light.

Condensation will occur when water evaporates and sits on the inside of the glass. So if you situate your terrarium in a sunny spot, the side of the container where the light hits will usually form condensation because the water in the substrate evaporates more quickly. Condensation is not a cause for concern, and is usually only an aesthetic issue.

If the sight of it bothers you, wipe down the inside of the glass with folded paper towel or a microfiber cloth and water; if the opening to the container is small, some rolled-up paper towel on the end of a pair of tweezers does the trick. I never use any chemical products to clean the inside of planted terrariums.

OPPOSITE I like to let dead leaves decompose in the terrarium. If springtails are added they will feed on these, but even in a nonbioactive terrarium I often leave them; there is a kind of beauty in imperfection.

Light

My inbox is filled with questions from people who are curious about why their terrariums aren't thriving, and when I ask for a picture of where the terrariums sit, more often than not the problem is revealed: a lack of light. That's usually because the right place aesthetically speaking is often a poor choice for the health of your plants.

Why plants need light

There is a misconception that terrarium plants and mosses can thrive in shady areas of your home, but in fact these plants need more light than you may think.

All plants use light to photosynthesize, which is the process by which plants convert sunlight into chemical energy to fuel their growth, and terrarium plants are no exception. The chlorophyll in plant cells absorbs light energy, which is then used to produce glucose and oxygen from the carbon dioxide and water that the plant takes in. This glucose serves as the primary source of energy for plant metabolism and growth. So, essentially, plants need a source of light to be able to make their own food.

RIGHT *Philodendron* 'Mini.' A rare, tiny-leaved species that is perfect for any kind of terrarium.

How much light does a terrarium need?

As long as my terrariums get a minimum of 12 hours of light in a day, I'm happy. I don't set timers for this, but you can if you like. I simply turn on the grow lights when I wake up and off when I go to sleep (roughly a 17-hour cycle) and I've had no issues with any plants on this setting.

However, time periods and quantities of light do vary, as there are often many different plants growing together in one terrarium and these various species require differing light intensities. For example, forest-floor–dwelling jewel orchids and *Selaginella* need lower light levels and will discolor in brighter light conditions, but *Dicranum scoparium*, a terrestrial moss, needs high light levels to survive or it becomes etiolated (growing long and scraggy due to lack of light) and eventually dies. Making educated choices about which plants we place together is important for their survival.

Even though terrariums often contain numerous kinds of plants and mosses, all with different growing needs, I find it helpful to view them as a single entity. In a confined space it's not possible to give every plant the precise care they individually require, and naturally some plants will perform better than others. Some are stronger than others, some weaker, and they will all compete for nutrients and light. While plants need light in order to photosynthesize, it is important not to leave lights on 24 hours a day, as they also need time in darkness to regenerate a key compound called phytochrome. This plays an important role in photoperiodism, which is the physiological response that plants make to varying lengths of day and night. This light-sensitive protein acts as a molecular switch that enables plants to sense and to measure day length and to then determine appropriate times for developmental processes like flowering. In photoperiodism, this ensures that plants flower at the most opportune times and maximizes their chances of reproductive success. Without phytochrome's role in sensing and responding to changes in light conditions, plants would lack the necessary cues to align their life cycles with the changing seasons.

By measuring the length of darkness, plants can decide when to bloom, when to enter dormancy, or when to undergo other important changes. During periods of darkness, plants rely on respiration to break down their stored carbohydrates and to release energy needed for growth and metabolic activities.

Darkness is also crucial for regulating hormone levels in plants, allowing them to respond properly to the world around them. One important hormone called auxin is responsible for things like elongating cells,

LEFT The stems of this *Nephrolepsis* fern are showing signs of etiolation (growing long and scraggy), which are caused by a lack of light.

promoting growth, and guiding directional growth responses.

Darkness gives plants a chance to rest and recover and to conserve energy by slowing down their metabolic activity. This energy conservation is important for their overall health and sustainability and allows them to repair any damage that may have happened during the day.

Of course, different plants have different needs when it comes to darkness, as much as they do for light. Some plants, such as cacti and succulents, can handle longer periods of darkness without any issues, while plants classified as long-day or short-day have specific requirements for the length of darkness in order to trigger specific stages of growth.

For example, spinach (*Spinacia oleracea*) is considered a long-day plant, as it requires a longer duration of daylight in order to start flowering. When exposed to shorter periods of darkness and longer periods of light, spinach tend to produce more leaves. Chrysanthemums (*Chrysanthemum* sp.), on the other hand, are short-day plants, and as such they require a shorter duration of daylight to start flowering. When the length of darkness exceeds a specific threshold, chrysanthemums initiate their flowering phase, so they often bloom in response to shorter days during fall or winter.

Are my plants getting enough light?

Signs your terrarium is getting too much light:

- the plants appear crispy
- the leaves have turned pale and the plants appear unhealthy
- there is an excessive amount of condensation on the glass
- the glass container is hot to the touch
- heat escapes the terrarium when you open it

Signs your terrarium is getting too little light:

- the plants and mosses etiolate (grow leggy in search of light) and the space between the leaf nodes is large
- the plants are dying
- usually small-leaved species have grown larger leaves
- mold is growing

Maximize success by measuring light

While many terrarium plants grow on the forest floor and understory in their natural habitats, what's considered low light outside is still considerably brighter than what we may think is a low-light area in our home. So we need to measure the light we're working with before we position terrrariums around our homes.

To remove any ambiguity around how much light your plant needs, get yourself a light meter—they are inexpensive and will help you to determine the perfect spot to place your terrarium. These devices measure how many foot-candles (fc) of light the space receives; this is a unit of measurement that quantifies the amount of light falling on a surface. Various plant species require vastly different amounts of light to survive, but as a general rule, many terrarium plants grow happily within the 150–250fc range. For my terrariums, I aim for no less than 150fc and have had no real issues with this.

Always measure the amount of light from where the plants are in your terrarium, not the top of the container—the light intensity will reduce substantially the further away the plants are from the source. There are also free apps on mobile devices that do the same thing, and while they aren't as accurate as a light meter, they can give you a decent idea as to how much light an area is getting.

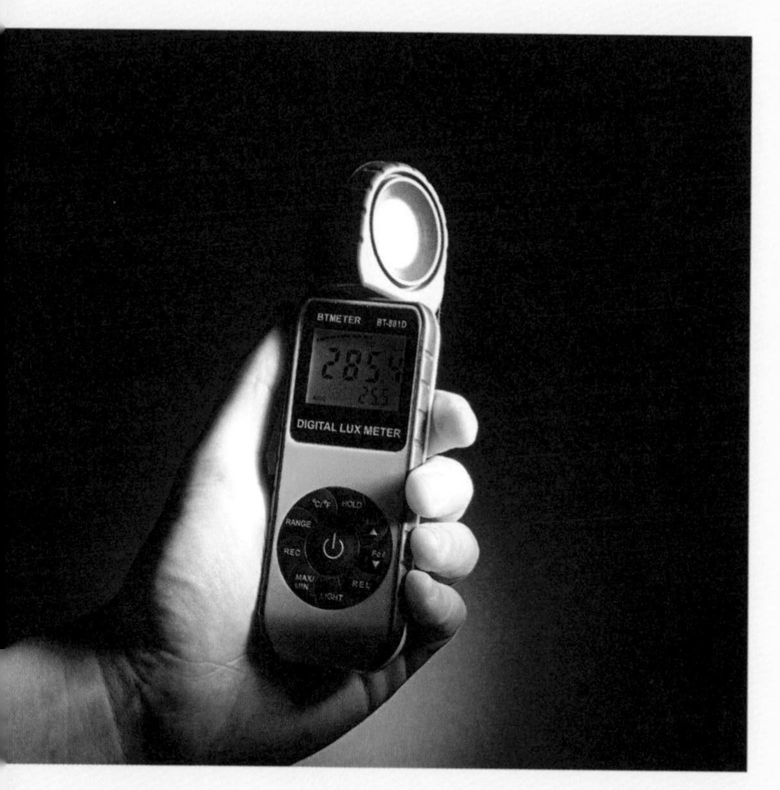

LEFT A light meter is a nonambiguous way of deciding the best spot for your terrarium.

Sources of light

We know that light is a vital component of plant health and growth, and there are numerous ways you can get the optimum levels for your plants, whether naturally or artificially.

NATURAL

It's hard to beat natural light, and the plants and terrariums I've grown this way almost always perform better than those under artificial lights. However, when choosing a growing position in your home you do need to consider the seasons as well as the aspect of your window—the direction is facing. The information about aspects here is for those in the northern hemisphere; if you are in the southern hemisphere, reverse the information—south for north, east for west.

The seasons A little morning or evening sun every day can be good for terrariums, but in spring, fall, and winter the sun is significantly less intense than in summer. So as the sun moves across the sky throughout the year, observe how the light hitting your window changes, then alter the position of your terrariums accordingly. Always be cautious when leaving your terrarium in the sun for any length of time, and if you're unsure, move it further away from the window to reduce the intensity of the sunlight. During the winter, move terrariums away from cold windowsills and don't sit them above radiators—while these will not sap moisture from the atmosphere in a terrarium, they can heat them up too much.

North-facing This is a good option because windows that face north receive no direct sun but still allow the terrarium to access a decent amount of light. I've found plants tend to respond well in this environment. It's worth noting that a cloudy sky through a north-facing window will provide much less light than a clear sky on a bright day. In my north-facing window, a clear sky on a bright day will produce upward of 400fc of light. Even tender plants will enjoy this. However, on a cloudy day, the window will receive 200fc of light.

South-facing A total no-go for terrariums, due to the length and intensity of light. A terrarium is a closed environment, so you are creating a greenhouse effect—and if you've ever walked into a greenhouse on a hot day, you'll know just how uncomfortable this can be! In this situation, your terrarium effectively becomes a mini oven, cooking everything inside it.

West-facing The sun is at its hottest at the back end of the day, so I avoid placing terrariums on west-facing windows during the spring and summer months. On a west-facing window, behind the clouds the light can sit in the 400fc range, but when it comes out, the sun will produce 800fc+, which can present a problem.

East-facing A good option for most of the year. On hot summer days it's worth moving terrariums away from the window, as even a little morning sun on hot days can be too much. Plants such as jewel orchids and *Selaginella* won't like being in direct sun for any amount of time.

Net curtains and frosted windows
Sometimes blocking the amount of light coming through a window can be beneficial for plants growing in a terrarium. If you have windows that face in a southerly direction, a net curtain of some kind is a good idea. In the bathroom of my house, which faces south, I have a frosted window that on a sunny day gets 200fc of light. I've been successfully growing a *Phalaenopsis* orchid and a *Nephrolepis* fern there for the past year when usually a south-facing window wouldn't be a suitable place for them. The frosted window reduces the intensity of the sun and creates an optimal condition for the plants.

ARTIFICIAL LIGHT

Using artificial light sources enables you to place a terrarium anywhere in your home, and many terrariums now come with built-in lights, which means that those otherwise unsuitable dark shady corners are now available for growing!

There are a variety of options, from simple warm/daylight/cool white bulbs placed in regular lamps, to high-end, full-spectrum lighting that is specifically designed for aquariums and vivariums and will work in your terrariums. The majority of my terrariums sit under artificial lighting; the increase in electrical costs is something to consider if you choose this option, but LED lights are effective. I use striplights, but find that one isn't enough, so I place two or three close together. There are different lighting colors available, but I favor the warm, white LED striplights as I think they look sleek, and they also don't produce much heat and provide a nice color.

This leads nicely on to the story about my oldest terrarium.

RIGHT Under-counter striplights can be attached underneath a shelf above your terrariums.

OPPOSITE This Arcadia Jungle Dawn LED bar has a high light output, so it must be positioned further away from the plants. This more expensive option is good for large vivariums.

My oldest terrarium

I remember picking up the demijohn bottle I used to make this terrarium from a local Facebook seller in 2019. I tried haggling him down from £5 (around $6), but he was having none of it.

When I planted this up, I added a layer of lava rock on top of the growing medium as an experiment, but it led me to a discovery. By placing a layer of porous, fine-grained rock on the soil surface as a top-dressing, shallow-rooted plants, like many kinds of *Peperomia*, or epiphytes, thrive on it. This is because they dislike sitting in constantly wet soil and prefer it to dry out between waterings. Terrarium substrate is naturally well draining, but the layer of porous rock creates a few centimeters where the roots can sit in a damp but airy space.

Even though this is a seemingly minor discovery, I now consistently implement this when I make terrariums. If I'm not using a bonsai substrate, I like to top-dress any open areas of soil, as this slows moisture loss, and if a degradable material is used it will also nourish the soil.

This terrarium has mainly been situated underneath two warm, white, LED striplights (not intended for plant growth but perfectly suitable), but has also had a period of time where it sat in a north-east–facing window.

Over the four years in which it has been planted, like me it has undergone a variety of changes. I initially planted it with some *Peperomia* 'Pepperspot,' some *Pyrrosia nummularifolia*, and *Ficus thunbergii* (also known as *Ficus quercifolia*—the oak leaf fig). However, the dominant plant is now the *Ficus thunbergii,* which has taken over the whole container. The *Peperomia* died off after a few years as the *Ficus* hogged all the light, and I removed the *Pyrrosia* fern because I didn't want the same thing to happen to it!

As time goes on, and the slow-growing *Ficus thunbergii* makes its way up the bottle, the leaves at the bottom and in the center can become starved of light. I counter this by occasionally pruning the plants and using the cuttings in other terrariums. This is one of the most rewarding parts of this terrarium; this specimen has become one large mother plant of *Ficus thunbergii,* and as long as I don't prune it too often, I have a near-endless supply of this beautiful plant.

When I created this terrarium, I was still an amateur hobbyist working as a mail carrier and trying to cram in as much time with my plants as possible. At that time, my social media presence was still very much in its infancy, and I hadn't started my YouTube channel. So much has happened in those four years; I've lost relationships but gained new ones, too, and seeing this terrarium on a daily basis reminds me of a bit of wisdom I heard, that you can't make any gardening mistakes, they are all experiments.

RUM

2

Terrariums to get started

What you need to build a terrarium

Investing in a good set of tools means having equipment that should last you a lifetime. It has been ten years since I got my first bonsai tools, and they're just as good today as they were when I bought them. Thankfully, you don't need a huge list of equipment to get started in a terrarium hobby, a few basics will set you up nicely.

CONTAINERS

One of the joys of this hobby is that it is a fun way to upcycle containers, and you can involve children in the process. You can reuse almost any containers that would be destined for the trash can and turn them into vibrant little ecosystems! I've had success repurposing jam and jelly jars, candy jars, spice jars, candy containers, cheese-ball containers, sauce bottles, soda bottles... You can see where I'm going with this!

There is also a wide range of containers that are specifically built for use as a terrarium. Small containers such as those built by Bioloark have air holes in the side to ensure the environment doesn't become stagnant, or you can get egg-shaped containers that have a light built into the top, which can be placed anywhere in your home. Going up a few more sizes are biOrb, Exo Terra, and In Situ terrariums; all are designed to house amphibians and reptiles, so they are perfect for use as vivariums or paludariums, but they can also be used just for plants, if you prefer.

Going up another level, you can even use glass cabinets. IKEA greenhouse terrariums are popular right now, and these short and tall cabinets have been turned into lush wonderlands housing lots of rare and unusual plants. So much so that there is now an Instagram page dedicated to them!

Colored glass can work but it does compromise how much of the contents are visible, and it also diminishes the amount of light that can get through. By all means try it out, but I recommend using clear glass.

Whatever kind of container you use, make sure it has a lid of some kind. The lid keeps the humidity inside, which is precisely what makes a terrarium so special. Open containers don't function well as a terrarium, unless the openings are very small.

ABOVE From miniature spirit bottles to high-end glassware with built-in lights, when it comes to terrariums, anything goes.

LIDS

Finding a container that is perfect for a terrarium but has no lid can be frustrating, but there is a little secret to solve this that I can share with you. Acrylic companies manufacture clear disks that can be cut to size to place over the top of your container, which opens up the possibility of using many more types of containers (see page 89)!

Where possible, do opt for a clear lid, because while opaque lids are fine as long as the opening is small, or if the light is coming in from the side of the terrarium, a large opaque lid will block out a lot of light and look strange.

DRAINAGE MATERIAL

This layer can be made up of any kind of gravel, stone, or pebbles, and its function is to allow water to drain away from the substrate. In smaller terrariums it's not necessary to have a drainage layer as long as you follow correct watering techniques, but in larger builds it does play an important role. If water does start to pool in this layer, that's a good indication that you've overwatered (see page 34).

In almost all circumstances where I use a drainage layer, I opt for expanded clay pebbles (often referred to as LECA, see page 27) as they come in a range of sizes and add next to no weight to the terrarium. This is especially important when making a large terrarium, as the amount of gravel or stones needed can make the container very heavy. My personal choice is to opt for a small-grained drainage layer in most builds, but larger grains in bigger terrariums. Smaller grains have better water-wicking abilities, but larger grains have more space for air gaps. In addition, the drainage layer works like a small air pocket at the bottom of the terrarium, helping to create an aerobic environment that is beneficial to the plants. As time goes on you'll see the plants' roots sitting happily in here.

MESH

Adding a mesh on top of the drainage layer prevents the substrate falling into it. In small terrariums, or those where you've used a fine-grain material as a drainage layer, you can skip using mesh altogether, provided you are careful with watering.

You can use a range of materials for this layer, but whatever you decide on, it's important to choose one that will not degrade over time.

Sphagnum moss is a suitable option for a mesh layer as it serves a dual purpose: plants love rooting into it while it also very effectively stops the substrate falling into the drainage layer.

Window mesh is also a good option, but choose a plastic one rather than metal as it doesn't rust. Weed-suppressant membrane is a popular choice, with some having a life-span guarantee of 25 years (if your terrarium lasts 25 years, congratulations!). If some substrate does fall into the drainage layer, it's not the end of the world. If you're using small grains for drainage you can skip the mesh altogether (just be careful when watering).

SUBSTRATE

In a terrarium, the substrate holds on to water, provides sustenance and a secure home for the plants' roots, and as such it's a fundamental aspect of plant growth. Terrarium substrates are made up of many different components and the size of the terrarium and plant choices determine which substrate is best to use, so refer to Chapter 1 when deciding which substrate to use.

Recommended drainage materials

Some of my favorite drainage layer options are:

- Moler clay (see page 25)
- Expanded clay pebbles (see page 27)
- Pea gravel
- Bonsai mediums (see pages 24–30)
- Soil Ninja Semi-Hydro Mix

Useful tools for creating terrariums

Long-handled tweezers Used to position cuttings in hard-to-reach places. It's worth spending a little more on these as the cheap ones tend to break after a short time. Tropica or ADA are my favored choices

Long-handled scissors An important tool in the toolbox. Very helpful when pruning, especially in containers with small openings

Short-handled scissors More durable than long-handled scissors, these are often used to cut plants with woodier growth

Chopsticks Having a few chopsticks in a range of sizes is invaluable; they're great for firming plants in place, especially in large terrariums or terrariums with small openings

Funnels Useful for adding substrate into terrariums with small openings

Scoops Good for scooping soils and soil components

Microfiber cloth and brushes For cleaning the glass or clearing away dust and debris

Aesthetic principles

I'll always remember something that a gardener I worked with once told me; I was mowing a path in some long grass when I came across a beautiful statue. I asked why it was placed here and not somewhere that people could always see it, and she responded by saying: "Not everything needs to be visible all at once."

How does this relate to gardening or terrarium building? We don't need to reveal everything, all at once, in every terrarium we make. Musician Miles Davis said: "It's not about the notes you play, but the space you leave." This is true in so many fields.

I believe it should be a gardener's aim to create a sense of cohesion and balance within the landscaping and planting. The distribution of elements is open to interpretation, but there are some guiding principles that are useful. Formal gardens often use symmetry and uniformly clipped plants to create a neat and tidy aesthetic, while naturalistic gardens have relaxed rules and favor asymmetry, allowing for a freer growth of plants. Whatever the style, a unified garden design ensures that all of the elements work together harmoniously. So think of the terrarium as a garden; and remember that the plants and elements within it should be proportionate to the size of the space—a small container with a large-leaved plant just isn't going to look right.

While some rules and guidelines are useful, following these to the letter every time will lead to common-looking spaces or, even worse, boring ones. It has been said on many occasion that in gardening there are no rules, only experiments, which is why contrast and a little bending of the rules often leads to new and exciting discoveries.

Whether your garden spans acres and is peppered with wildflowers and long grasses or your garden exists in a glass jar that relies on the power of your desk lamp for its very existence, we are all gardeners, and the privilege of working with plants is something we should never take for granted.

OPPOSITE The terrarium on the left is an example of how I would use colorful plants in a terrarium; the two others are naturalistic terrariums where I feel colorful plants would spoil the aesthetic.

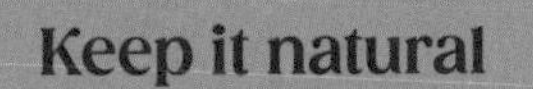

Keep it natural

I think that really colorful plants, such as bright-pink *Fittonia*, don't fit the aesthetic of a naturalistic terrarium. I'm not saying you should avoid them, because I do sometimes use them in little groups (see the Colorful Fittonia Terrarium guide on page 177), but placing a bright-pink plant in the center of a terrarium that consists of beautiful green mosses and plants, along with the natural dark-brown hues from decaying branches, can look rather strange. I do think colorful plants have their place, just not in naturalistic-looking terrariums.

Everything has a front

There have been many lessons I've learned in my horticultural journey, but the one I seem to use most is that everything has a side from which it looks best. In bonsai, a tree is viewed from its best angle. A piece of Seiryu stone may have a totally flat side and one with beautiful patterns on it. A branch may have one side that has lots of texture and one that is totally smooth. A plant may have its leaves tilted in one direction, so it looks better having that side facing forward. Try to find the best-looking side to the plant or object you're using, or adjust it to get the look you want, then make sure that's the one that is showcased in your design layout.

Background, midground, and foreground

Using background, midground, and foreground plants along with pieces of hardscape in a terrarium has both aesthetic and functional purposes. The layering of different plant types adds depth and visual interest, often mimicking the arrangements found in nature. The layering effect within a terrarium introduces complexity to the landscape, making it look more believable and aesthetically pleasing.

Plants come in varying sizes and should be positioned accordingly in a terrarium. Taller plants are usually used at the back of the terrarium so they don't block anything behind them; midground plants are smaller and are placed in front of the background plants, while the smallest plants should sit at the front of the container. Common foreground plants include mosses and creeping plants that grow horizontally rather than vertically.

Arranging plants in this manner manages the space inside the terrarium more efficiently and creates a sense of balance and harmony. The different layers will complement, rather than combat, each other.

Again, these rules are there simply to guide us, and breaking them in creative ways opens up new compositions. For example, small-leaved background plants can be strategically positioned in the terrarium to climb a backdrop feature, or placing a midground plant at the front or back can be an effective way of creating a natural feel. For a more dramatic effect, placing a tall background plant further forward can make the landscape within the terrarium look more believable! I like to place creeping plants in between the mid- and background plants, too, because as they grow it creates a pleasant surprise when they pop up seemingly out of nowhere!

Avoiding symmetry

Symmetry is often found in nature—snowflakes and sunflowers are perfectly symmetrical—but if you take a look at a natural landscape you'll see that symmetry is rare, and if it is present it usually indicates the influence of humans. How you plant and hardscape a terrarium is totally down to you, but in my opinion, favoring odd numbers and avoiding symmetry always looks more natural.

When there are an even number of elements, the eye naturally pairs them up, which creates a sense of balance and completeness. This can be pleasing, but I've found it feels static when utilized in a terrarium. Odd numbers of elements, on the other hand, create a sense of dynamic imbalance that is more engaging and visually interesting. It also allows for a center of interest or a focal point. This is particularly true for groups of three, where there is a natural middle as the central element draws the eye and becomes a focal point around which other elements are arranged.

OPPOSITE A good example of an upcycling project. The pickle jar is from a fish-and-chip shop and the moss from my mother's garden. The *Ficus thunbergii* started as a small cutting.

LEFT Hardscape materials and plants can look more natural when positioned in accordance with the Rule of Thirds.

Planting using the Rule of Thirds

The Rule of Thirds is a compositional guide that is used in photography, the visual arts, and design to create balanced and visually appealing compositions. It involves dividing an image into nine equal parts by overlaying two equally spaced horizontal lines and two equally spaced vertical lines, which results in a grid with four intersection points.

The key idea behind the Rule of Thirds is to avoid placing the main subject of the image at the center of the piece, instead aiming to position it along one of the intersecting points of the grid. In addition, key elements, such as particularly nice hardscaping or a special plant, should be positioned along the lines themselves. This placement creates a more dynamic and interesting composition, which draws the viewer's eye through the image and encourages a sense of movement and of balance.

While the Rule of Thirds is a helpful guideline for achieving well-composed images, remember that terrarium-making is subjective, so this is not a strict rule to follow, rather it's a tool that creators can choose to follow or deviate from, based on the vision they wish to convey through their work. So while following this rule can help you create impactful and aesthetically pleasing terrariums, it's important to embrace your own creativity and individual expression, too.

Letting nature do its thing

If you have an expectation that over time your terrarium will continue to look the same as it did on the day it was planted, you will be disappointed. I have come to adore the way an overgrown terrarium looks! It may not be a beautifully curated landscape, but there is something almost magical about it.

Allowing plants to grow without too much restriction is important, as it allows them to develop vigor. And while the occasional pruning is necessary to maintain a plant's well-being and shape—especially to prevent plants overwhelming others in the terrarium—excessive pruning can negatively impact their growth. For instance, plants need their leaves for photosynthesis, which gives them energy (see page 40), but regular pruning removes these crucial leaves, causing weakened growth and stress on the plant. Over-pruned plants also become more susceptible to pests and diseases, because their natural defense mechanisms weaken. Imbalanced growth, depleted energy reserves, and a loss of aesthetic appeal are other negative consequences.

To avoid these issues, it's important to follow appropriate pruning practices based on each plants' specific needs (see Chapter 5). Pruning should be selective and purposeful; don't just prune to keep the plant looking the way it did on the day you planted it; prune to remove damaged or dying leaves, or if the plant is too big and blocking out light for anything underneath.

How to create maximum impact with minimum cost

There is a reason why certain plants are more popular in terrariums than others, and it's often because they are so reliably effective. However, if we only use these plants in a conventional way, our terrariums will look generic and mass produced.

While I love using rare plants—and some of the varieties available are stunning —the more common plants can be just as impactful if you know how to use them. They are also significantly cheaper.

Taking cuttings from these plants (see page 206) has a dual purpose for terrarium hobbyists. First, it is an excellent way to get the scale of plants right for your terrarium space, because you can choose just a tiny part of the plant rather than try to squash in the whole thing. Second, it's a really financially effective way to increase the amount of plants you have to work with. You can buy just one plant and get multiple plants from it, or you can swap cuttings with other enthusiasts. In addition, you can choose the part of the plant that's best suited to your design for maximum impact. For example, new growth on a *Fittonia* looks beautiful and dainty, and *Peperomia verticillata* has stunning juvenile growth.

Cuttings will stay smaller for longer while they establish a root system, which is another benefit for the confined space— less maintenance! It's important to allow the plants to grow so they can develop vigor, before pruning back overenthusiastic plants to fit the space. You can also prune strategically to get the cuttings to bush out rather than grow upward, and produce a more compact shape. If you have any plant failures, you can simply replace specimens with more cuttings from the mother plant.

The influence of bonsai

Bonsai and other Asian gardening influences have had a significant impact on my horticultural journey, and I've found that many of the skills I learned through bonsai almost directly transfer over to other forms of horticulture, including terrarium making.

One of the standout features of bonsai is the practice of strategically pruning trees to grow in a certain way. This takes a lot of skill and is species-specific; generally speaking, in a terrarium the plants we use aren't as fussy about their pruning requirements, and simply pruning them before they get too big, with an emphasis on keeping them smaller and bushier, goes a long way.

Here are some key bonsai styles that I think are most relevant to the scale and growing environment of terrariums.

LITERATI

Of all the forms of bonsai, nothing captured my heart more than the *literati/bunjin* style. The original meaning of the term *bunjin* was not related to a particular bonsai style as it is today, but used to reference any bonsai that was admired or valued by learned men (*bunjin*). It is only since the nineteenth century that the term *literati* has been used to refer to the stylistic features that this style is known for today.

It's believed that the Chinese Southern Song paintings are the inspiration for *literati*-style bonsai. The paintings feature tall, slender trees with minimal taper and sparse foliage on the upper third of the trunk.

Over the years, *literati* trees have changed in style and appearance, maintaining the essence of the original style with a slender, nontapered trunk, and sparse foliage, but also incorporating dramatic movement in the trunk with a much more contorted appearance.

The essence of the *literati* form is to find beauty in simplicity, and this can be translated to terrariums by using dramatic lines, minimal foliage, and some cleverly positioned hardscape.

WABI-SABI

This Japanese aesthetic concept invites us to find beauty in the fleeting moments, to appreciate the imperfect aspects of ourselves and others, and to embrace the natural cycles of growth and decay. It encourages us to let go of the pursuit of idealized standards and instead embrace the beauty that lies in the uniqueness and impermanence of the world around us.

"*Wabi*" refers to the rustic beauty and simplicity found in nature and objects, often associated with modesty, austerity, and solitude, and is often described as unpretentious and unadorned. "*Sabi*" represents the beauty that comes with the passage of time, the patina of age, and the signs of wear and use.

Wabi-sabi is not about seeking perfection or clinging on to material possessions; it encourages an awareness and mindfulness of the present moment, to find beauty in the ordinary and the imperfect. It values the inherent characteristics of materials, such as the texture of wood, the cracks in pottery, or the weathered appearance of stone. It emphasizes the simplicity and subtlety of design, avoiding excessive ornamentation and embracing a sense of harmony and tranquility.

Applying the ideas of *wabi-sabi* to terrarium making is extremely useful. Plants are perfect and imperfect at the same time. As plants mature, older leaves will naturally die off once they've fulfilled their role of providing the plant with food through photosynthesis. I glance at the *Chirita tamiana* in my frog tank that has flowered nonstop for the past eleven months. It's looking a little sad as the other plants have grown and blocked out some light to it, so the plant is on its way out, but there is a gentle beauty that it exudes as it lives out its final days. It has lived its life, producing many blooms along the way, but soon it'll be one with the soil, the cleanup crew will consume it, then the remnants will provide nourishment for the other plants in the terrarium.

Wabi-sabi has taught me to accept that everything is impermanent, everything has a beginning and an end, and this has helped me immensely on my terrarium journey and stopped me from being upset when my efforts didn't work out. And now, as Darryl Cheng, author of *The New Plant Parent*, reminds us to, I thank each leaf for its service.

PENJING

This traditional Chinese art form involves the cultivation of miniature landscapes in containers. The word *penjing* combines two Chinese characters: "*pen,*" meaning "tray" or "pot," and "*jing,*" meaning "scenery" or "landscape." This form of bonsai aims to depict a natural scene, like a mountain, forest, or river, in a small and artistic representation. Carefully selected trees, shrubs, and other plants mimic the grandeur of nature on a reduced scale, while hardscape materials such as stones, pebbles, and sand replicate mountains and cliffs, with rivers within the landscapes completing the scene. Pot selection and scale are of the utmost importance in this style.

As a living art form, *penjing* requires ongoing care and attention to maintain the health and aesthetic appeal of the miniature landscapes. It is not just a visual art but also an embodiment of the artist's appreciation for the beauty of nature and a reflection of the harmony between humans and the natural world.

Simple terrarium in a fishbowl

MATERIALS

Plastic wire window mesh (0.28mm holes)
Clear glass fishbowl (10in/25cm in diameter)
Permanent marker pen
Scissors
Expanded clay pebbles
Terrarium substrate (simple or premium premix works, see page 31)
Pressurized spray bottle filled with distilled or deionized water
Hardscape items: cork bark (or use lava rocks or grape wood)
Bowl filled with 1¾ pints (1 liter) of water
Lid, at least 3mm thick (see page 53)
Long-handled scissors

PLANTS

Nephrolepis ferns
Leucobryum glaucum (moss)
Pilea glauca
Ficus pumila
Fittonia albivenis

This project is the perfect starter terrarium because the bowl has a large opening that makes assembly easy, but also because of the simplicity of the hardscape and the toughness of the plants used here. The only prerequisite for the container is that it's clear and transparent, so light can penetrate. Glass is a good choice, as it is strong and easy to clean, and, unlike plastic, it rarely scratches and doesn't dull over time.

Cork bark makes an excellent hardscape material, as it is naturally textured and adds detail to the terrarium. *Nephrolepis* ferns occupy the background due to their size and ability to fill empty spaces in the arrangement, while cuttings taken from plants of *Pilea glauca, Ficus pumila,* and *Fittonia albivenis* are dotted in and around the *Leucobryum glaucum* moss and cork bark as the finishing touches.

CONTINUED ⟶

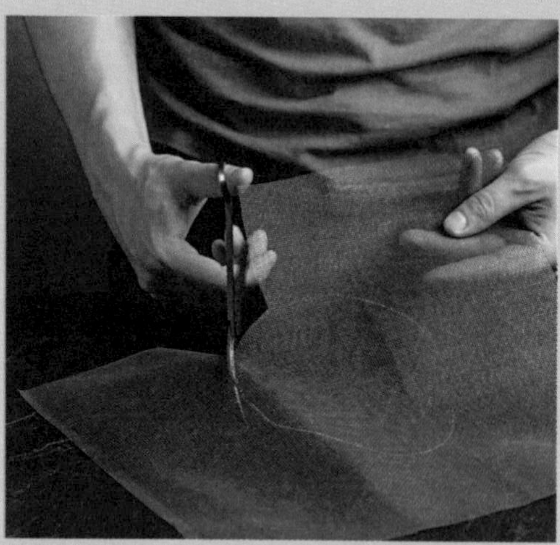

1. Lay the piece of mesh on a sturdy surface and place the container on top. Draw around the base of the bowl on the window mesh with a permanent marker, then cut the piece to size and set aside.

2. Place a 2in (5cm) layer of expanded clay pebbles in the base of the container and cover with the window mesh—don't worry if there is a little overhang or it's slightly too small.

3. Scatter the substrate over the mesh, arranging it on a gradient so it's higher at the back and lower at the front. This allows more surface area to be visible from the front. Spray the substrate until evenly damp—this helps to keep it in place when adding the hardscape and plants.

4. Place your pieces of cork bark into the bowl and firm into the substrate. Odd numbers and nonsymmetrical patterns look more natural, but how you arrange them is totally up to you—just be sure to leave some space behind it to fit in the fern.

5. Take the fern out of its pot and divide the root ball in two. Using your fingers, find the center point between the leaves, then, from the top, carefully but firmly split the leaves and root ball into two even pieces, leaving as much root intact as possible. You should be left with two plants—some leaves will inevitably be lost during this process, but that's nothing to worry about. Gently loosen the root ball and remove a little soil but try to keep the roots as intact as possible; this helps the roots find their way into the new substrate.

6. At the back of the terrarium, behind the cork bark and slightly off-center, make two holes that are large enough for the fern root balls. Put each divided fern into a hole and firm in place. If any leaves are poking out of the top of the bowl, prune them at the base of their stems using the long-handled scissors.

7. Take a piece of *L. glaucum* moss and submerge it in the bowl of water. Give it a good squeeze until you see air bubbles. Remove the moss from the water and squeeze it again to remove as much water as possible. You should be left with a damp piece of moss.

8. Looking at the moss from the side, you'll see that it's divided into two colors—the lower beige part and the green at the top. Split the moss into smaller 2in (5cm) pieces and, using the scissors, trim the beige part until it's around ½in (1cm) beneath the green part.

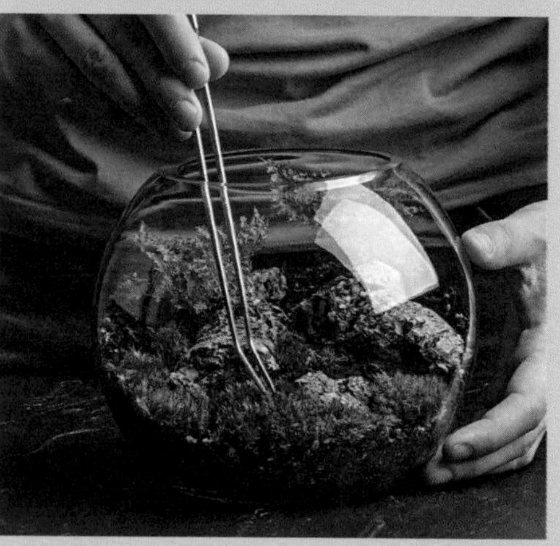

9. Place the moss around the front of the terrarium (the lowest part of the substrate), and around the cork bark. Don't worry about being overly neat at this stage as we will soften any lines with the plant cuttings.

10. Take the *Pilea glauca*, *Ficus pumila,* and *Fittonia albivenis* and some rootless cuttings, favoring the newer growth with smaller leaves, and place them into the moss and around the open areas of the terrarium. Don't overfill the bowl, as the plants will grow over time and fill space.

11. Finally, top with the lid and place your terrarium somewhere light and bright, or position it under a grow light.

Terrarium in an aquarium

MATERIALS

Fish tank with a light and a closed lid
Scissors
Terrarium substrate (simple, premium, or premix, see page 31)
Driftwood
Long-handled scissors
Long-handled tweezers
Pressurized spray bottle filled with distilled or deionized water
Bonsai medium or bark (see pages 24–30), for top-dressing
Lava rocks

PLANTS

Peperomia prostrata
Nephrolepis exaltata (Boston fern)
Taxiphyllum barbieri (Java moss)

This container makes a simple project, because the terrarium can be placed anywhere, as long as it's near an electrical outlet. Fish tanks are available in a range of sizes and prices, so choose one that fits your needs, space, and budget. I've chosen to use *Nephrolepis* ferns, as they add instant impact, and *Peperomia prostrata* cuttings and *Taxiphyllum barbieri,* which will grow perfectly on the bonsai medium or rock top-dressing.

CONTINUED ⟶

1. Add in the terrarium substrate until roughly one-fifth or a quarter of the container is filled. Arrange the substrate on a gradient so it's higher at the back and lower at the front. Don't be afraid to do this drastically! Lightly spray with water, then wipe down the sides of the terrarium.

2. Place the driftwood into the terrarium in an aesthetically pleasing way. Avoid placing anything right in the center of the terrarium.

3. Remove the *Nephrolepis* fern from its pot, loosen the root ball, and remove a little of the soil—but try to keep the roots as intact as possible.

4. Make holes in the substrate with your fingers where you want the plant to go, then firm it in place.

5. Top-dress the substrate with the bonsai medium or bark.

6. Take the Java moss and, with long-handled scissors, cut it into small pieces, roughly 1–3mm in size. Spread the cut pieces around the front of the terrarium, on top of the bonsai medium or bark. Lightly spray with water.

7. Then take rootless cuttings from the *Peperomia prostrata* and place them around the driftwood using long tweezers.

8. Add some lava rocks at the front of the terrarium.

9. Lightly spray the plants with water. Place the lid on and turn on the light.

Ongoing care

Prune the plants as needed. Any trimmings from the *Peperomia* plants can be used again, but the fern leaves should be discarded. While the cuttings establish, it can be helpful to lightly spray them every so often, as fish tanks vary widely in their conditions, with some keeping in lots of humidity and others holding very little.

Spirits bottle terrarium

MATERIALS

A clean spirits bottle with a lid
Activated charcoal
Terrarium substrate
Akadama or lava rock, for top-dressing
Pressurized spray bottle filled with distilled
  or deionized water
Long-handled tweezers
Scissors

PLANTS

Peperomia prostrata

I was surprised to discover that *Peperomia prostrata* is actually an epiphytic plant that comes from South American rainforests. At first glance, you'd be forgiven for thinking this is a succulent, with its plump leaves and preference for dry soil. When I placed some cuttings into a paludarium with a fogger, the leaves swelled and the cuttings quickly pushed out thick roots into the branch they were planted on.

I'd never seen it grow so healthily and it made me think, have I ever actually seen a healthy, mature *Peperomia prostrata* here in the UK? *Peperomia* love bright diffused light (minimum 150fc) and as much humidity as you can throw at them, which is difficult to achieve when they are grown as houseplants. They're often sold as hanging plants, but the medium they're grown in is far too water-retentive, which always leads to the leaves rotting away, so in my

CONTINUED ⟶

experience they are more successful in a terrarium. It's particularly important not to overwater *Peperomia*; keep the substrate slightly damp and the plants above the soil level. I advise opening the lid every few days to allow for some airflow, or, if you have a metal lid, punch a few holes in it using a hammer and nail. *Peperomia* has shallow roots and a preference for a well-drained substrate; top-dressing the surface with a fine bonsai medium creates a layer that stays damp but never wet—perfect for the cuttings to root into. This project uses a spirits bottle, so the surface area on which the plant can grow is small and in time the surface will be covered in a mass of succulent, circular leaves.

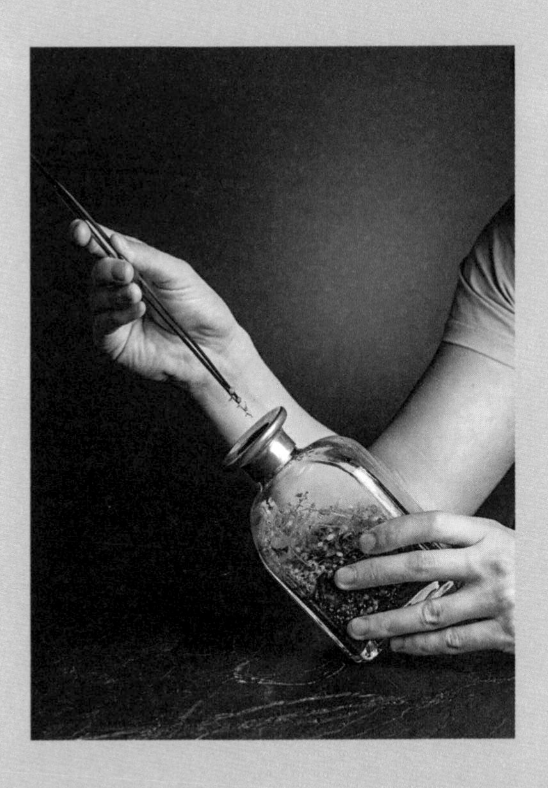

1. Place the bottle on a sturdy surface and remove the lid, then add a ½in (1cm) layer of activated charcoal into the base.

2. Fill a quarter of the bottle with terrarium substrate at a gradient so it's higher at the back and lower at the front, then add a ¾in (2cm) layer of akadama or lava rock on the surface of the substrate. Thoroughly water the substrate until it's evenly damp.

3. Select stems from your plants that have 2–5 leaves and remove as rootless cuttings. Gently lower the cuttings one by one into the bottle, using long-handled tweezers, until they rest on the surface of the substrate, ensuring the leaves are facing upward. It's easier to do this by tilting the jar.

4. Place the lid on the bottle.

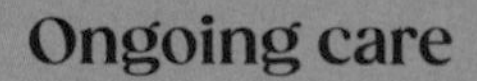

Ongoing care

The light will need to come through the front of the container, rather than the top, so take this into consideration when deciding where to place the terrarium. Only water as and when the substrate or moss starts to dry (see page 34), watering carefully using a spray bottle on the mist setting.

Once the cutting or the moss have outgrown the container, remove and replace with fresh plants.

Spice jar terrarium

MATERIALS

A clear glass spice jar
Dish soap
Cooking oil for removing stubborn glue
Funnel (optional)
Terrarium substrate (simple soil without chunky components)
Pressurized spray bottle filled with distilled or deionized water
Long-handled tweezers
Paper towel
Chopstick or pencil
A small piece of grape wood or cork bark

PLANTS

Leucobryum glaucum (moss)
Peperomia emarginella

My love for Indian cooking means I have an abundance of spice jars that are usually destined for the recycling can. In early 2023 I made a video where I turned an old spice jar into a terrarium using a little scraping of moss from the alley next to my home, which got millions of views across various social media platforms. It's a simple project that's best done using as attractive a spice jar as you can find.

CONTINUED ⟶

1. Clean the spice jar using warm water and dish soap. Peel off the label as best you can. If it leaves a residue, place that side of the jar onto a saucer of cooking oil for half an hour and scrape it off. Wash the container again using dish soap to remove any oil, then thoroughly dry the container.

2. Using the funnel, if you have one, pour the substrate into the jar. Fill just over a third or just under halfway up the jar. Lay the terrarium horizontally and tap the rim of the lid to coax the substrate along the jar, to achieve a gentle slope from bottom to neck. Spray the substrate with water until it's evenly damp, then, using tweezers and paper towel, carefully dry the visible glass so you can see what you're doing. Use a chopstick or pencil to gently firm in the substrate.

3. Cut pieces of moss to size. Using tweezers, place some moss upright at the bottom of the jar (the final terrarium will be upright, not horizontal). Gently press in place using a chopstick or pencil.

4. Break a piece of grape wood or cork bark to fit into the container, using your hands, and firm it into the substrate so that when you turn the container upright it stays in place. Fill around the grape wood/cork bark with more moss. You may find other species of moss growing within your moss; feel free to add that in too, as I've done!

5. Take a cutting from the *Peperomia emarginella* and gently firm it upright on top of the moss. Use the tweezers to put it in place. You can use another piece of moss to firm it in if needs be.

6. Sit the jar upright and place the lid on top and set in a bright position.

3

Expanding horizons

Personalizing terrariums

Creating a terrarium is a personal endeavor, and while there are a few guiding principles you should take into account, it's important that you experiment and figure out what works for you. The creative possibilities are vast in this hobby, and there are so many opportunities to imprint your own personal interests on it—from the different choices of hardscape materials, to the use of figurines and decorative items. Plus, of course, the near-endless variety of plants available.

Finding your own voice in your hobbies takes time and practice, and terrarium-making is no different. When I look back at my early social media posts I often cringe at my efforts and even contemplate deleting them. I never do, though, as I remind myself that they're there to show me how far I've come on this journey.

Figurines

Tiny figures can really personalize your terrarium, but finding the right ones for your project can be tricky. Model rail stores have a great selection but the majority of them are people with luggage or farm animals. The sheer volume of figurines on offer makes it difficult to find something suitable, but having a search through websites can mean discovering interesting pieces; I've found rock climbers, tigers, camels, people camping, zombies, and lots of other stuff!

The improvement of 3D printing technology in recent years has also helped to broaden the choice of figurines. I recently found an independent Etsy seller who stocked miniature painted dinosaurs. This naturally appealed to me, as I was obsessed with dinosaurs as a child, and no film gave me more sleepless nights than *Jurassic Park* (those velociraptors...). As someone who has loved playing video games from a young age, I was also excited to find some *Legend of Zelda* figurines online, which I used to recreate Kokiri Forest in a terrarium.

ABOVE A 1:87 figurine is the perfect size for a terrarium landscape.

I love that these figurines are detailed and come in a variety of scales, but the ones I use most for terrrariums are 1:87 (HO), 1:76 (00), and 1:148 (N gauge). Search online for the size you want and the right scale for your terrarium. Generally speaking, it's best to use plastic models, as metal can rust and wood will degrade in the high humidity, also potentially leaching unwanted chemicals into the ecosystem.

Figurines come in two options: painted and unpainted. If, like me, you paint like a toddler, it's better to purchase the painted models, but if you are proficient with a brush, painting your own is an excellent way to add a personal, unique touch to your terrarium scene. Acrylic paint is one of the best paints for water resistance; it is very durable and fast-drying, and will hold up in the humidity of a terrarium.

As a content creator, using figurines gives me a wonderful opportunity to make something different and also to make videos about games and movies that resonate with the people who sit outside of the terrarium niche. Just think about how many possibilities there are here…

Layering

It's common to see terrarium creators adding decorative layers to their builds, such as using different-colored sands to create wave effects and add texture, color, and interest to the parts of the terrarium below the substrate. While this is not something that I personally like to do—because I prefer all of the attention to be on the plants in my terrariums—it can look really effective.

LEFT *Specklinia dressleri* is a stunning, tiny-leaved orchid endemic to Panama.

Choosing and using plants

When people first start this hobby they are often unsure as to which plants to use, and may be limited to those they have already or that they know. But a little bit of research into suitable plants (see Chapter 5) can help you to pick some favorites.

I don't recommend beginners go out and buy rare plants, but to instead learn how to work with more common varieties such as *Fittonia*, *Ficus* sp., and *Pileas*. Once you get to grips with working with these plants, the same skills can be applied to the rarer varieties. Then as you progress through the hobby, you will naturally discover the more unusual plants that work in a terrarium. These always command a higher price point, but they are not always demanding in terms of care. For example, the micro orchid *Specklinia dressleri* (see page 187) is a rare species but it doesn't require much maintenance, and most Marcgravias (see page 169) are the easiest plants to care for—although the rarer varieties can be very expensive.

Some plants can be pricey, but for a terrarium you will mostly only need cuttings, so with a little propagation knowledge (see page 206) you can create many plants from just one. When choosing which plants to use for a terrarium, I consider the size of the leaves, growth habit, speed of growth, lighting requirements, soil preference, and the color and texture of foliage. We don't always know the answers to these questions but through a little experimentation we learn quickly what works and what doesn't.

This also applies to how you use plants in a terrarium. Instead of simply placing plants in a container, think about how you can creatively display them. For example, a creeping plant like a *Pyrrosia* fern or *Ficus thunbergii* will grow perfectly fine terrestrially, but if we position them so their new growth is facing a background or other hardscape item, it will grow onto it and attach its roots to the surface. Another option is to use a hollow piece of cork bark filled with sphagnum moss (see Fallen Branch project on page 106). This highly effective method creates the illusion the plants are growing epiphytically, when in fact the cork bark essentially becomes an elongated plant pot!

Unconventional containers

As eye-catching as a rare or unusually displayed plant can be, it can be taken to another level entirely through the use of an unexpected or unusual piece of glassware.

Opting for an unconventional container over a standard terrarium can really pique a viewer's interest. For instance, a while back, I saw that the terrarium creator Terrarium Designs had turned a Tic Tac box into a micro terrarium using mosses and tiny plants, which I thought was a genius idea!

This also goes to show that you can reuse pretty much any suitable container; old fishbowls, decorative glassware, bottles, and candy jars are all great options. I once reused an old, cork-topped saffron jar, bought at my local Asian grocers, which looked wonderful planted with a small amount of moss. I have also transformed larger, decorative jars, such as the sleek, clear container found in a homeware store, which I planted up with jewel orchids and *Selaginella* to incredible effect.

As long as your chosen container is clear, there are no limits to what you can use. Don't be afraid to try new things!

The hunt for interesting glassware

To find really unusual glassware for terrariums, look in unexpected places. Searching online is great, but nothing really beats seeing things in person, when you get a true sense of the condition, age, and size of the container. Here are some good places where I have found glassware that's a little out of the ordinary.

MARKETS AND YARD/GARAGE SALES

A day out to a local market or yard/garage sale can provide a great opportunity to find unique glassware at a good price, such as vintage containers and vases, Wardian cases, glass domes, or even demijohns and lab bottles. A bonus is that you can often haggle with vendors or owners to get a better price—never settle on the first price offered.

HOMEWARE AND DESIGNER STORES

If you're on a budget, check out the glassware sections of homeware shops. These often stock items such as candle holders, mason jars, glass food containers, and cloches that are not intended for use as a terrarium but actually work perfectly well! I've found unusual items in The Conran Shop, H&M, and Arket. Shop-bought glass containers are often sold without lids, but this is easily solved by measuring the opening and ordering a clear, acrylic disk to place on top.

ONLINE SELLERS

I've been lucky enough to find some really beautiful glassware on online marketplaces such as Etsy, eBay, Facebook Marketplace, and Amazon, but asking the right questions is a skill. Be creative with your online searches, rather than just searching for "terrarium glass," as this almost always results in mediocre, overpriced, and frankly boring glassware.

LOCAL GLASS BLOWERS

I once bought a set of glass orbs from the Belgian brand Serax. While not intended for terrarium use, the thin, small opening and the container's perfectly spherical shape make them ideal. I love them. The problem is, they discontinued the line, but I visited my local glass blower and asked the owner about recreating the orbs. Not only did they work out cheaper, but the glass was thicker and more robust. Each one was unique, too, with slight differences due to the construction process.

Globe terrarium

MATERIALS

4in (10cm) orb container
Funnel
Terrarium substrate
Pressurized spray bottle filled with distilled or deionized water
Long-handled tweezers
Chopstick or pencil
Tree fern fibers
Figurines
Small marble or acrylic lid to cover the opening

PLANTS

Leucobryum glaucum (moss)
Nephrolepis cordifolia 'Duffii'
Peperomia prostrata
Ficus 'Borneo Small'

There's something infinitely satisfying about holding a ball of plants in the palm of your hand. I first experimented with making these globe-style terrariums in 2019 and they were extremely popular on my Instagram page. Since then, I've refined the technique and plant choice, and I've added figurines like these hikers, too. I like to think of terrariums as miniature worlds that are magical and dreamlike in quality, and these characters are perfect. Mosses go a long way in creating a realistic scene for figurines, so I like to fully cover the substrate with a layer of *Leucobryum glaucum* moss. This also keeps the atmosphere humid, which is important in such a small terrarium. The *Nephrolepis* fern is divided into smaller pieces, which often reveal tiny "plantlets"; over time, the leaves need to be pruned to keep them to a size that will fit the container. *Peperomia prostrata*, meanwhile, is easy to keep. Simply rest these cuttings, leaves facing upward, on top of the moss, and in a short time they will push roots out through the moss and into the substrate.

CONTINUED ⟶

1. Place the orb on a sturdy surface and, using a funnel, add enough substrate to fill a quarter of the container, arranging it on a gradient so that it is higher at the back and lower at the front. Carefully spray the substrate with water until it is evenly damp.

2. Take small pieces of the moss and trim off most of the lower beige part. Fully cover the substrate with the green moss, lowering it into the orb using long-handled tweezers. If needs be, use the chopstick or pencil to dislodge the moss from the tweezers. Take your time.

3. Divide the fern into small pieces with both leaves and root attached—these should be small enough to fit into the orb. Carefully add 1 or 2 toward the back of the container using long-handled tweezers. Place the *Peperomia* and *Ficus* cuttings in and around the terrarium.

4. Finish off by positioning the tree fern fibers as accent pieces, place the figurines into the scene, then place the marble or lid on top of the opening.

Hardscape materials

The hardscape features you use are the difference between a terrarium looking interesting and *simply stunning*.

Nothing makes me happier than seeing a glass container filled with my favorite tropical plant species, growing on or around a characteristic branch or stone in a beautiful container. Hardscape materials have a practical purpose of adding structures like caves, arches, and tunnels to provide a surface for creeping plants to climb, but they also double up as a shelter for any custodians you might choose to add, such as isopods and millipedes. But they can also take a terrarium to a new level, aesthetically speaking. Hardscape offers the opportunity to introduce dynamic structures and unique features to your scheme, give a terrarium a naturalistic appearance, and help accentuate the featured plants.

Sterilizing items before use in a terrarium

It's commonplace in terrarium guides to see a section that tells you to sterilize your hardscape before use to get rid of any insect "hitchhikers." It's not something I've ever needed to do; buying materials from a reputable source instead of collecting your own from outside means this shouldn't be an issue. Instead, I give hardscape materials a brush to remove any dust and debris then soak them in water for a day or so. If I'm in a rush, I often skip this step and still have had no problems at all.

Rocks

Generally speaking, rocks are easy to add to a terrarium, with little to worry about in terms of them degrading or developing mold. Certain rocks do leach chemicals into the substrate, but that is more of an issue for aquascapers to worry about; I've never had any issues when adding rocks to a terrarium. When sourcing rocks, whether from nature or buying from suppliers, make sure they have not come from a potentially contaminated location, because rocks from a polluted stream or an industrial area can cause problems in your terrarium. Always consider the environmental impact of removing rocks from nature, for example, taking rocks or sand from beaches impacts erosion. When adding the rocks to your terrarium, secure them in place by carefully and firmly pressing them into the substrate after it has been watered.

LAVA ROCK

Also known as volcanic rock, lava rock is a popular choice for terrarium building, both as a soil addition (see page 25) and as a hardscape material. This intriguing rock is formed when gases in magma begin to solidify and harden, and it has a high porosity because of the thousands of pits and pockets created by the gas bubbles. Lava rock's ability to wick up water yet still remain airy makes it a good choice for growing creeping or epiphytic plants on.

This inert rock comes in a range of colors depending on its source, such as black, brick-red, light red, and deep orange. Larger lava rock chunks can be skilfully shaped by sandblasting to create stunning arches and rocks with holes through their centers.

SLATE AND SHALE

Shale is a sedimentary rock made of clay, quartz, and other minerals, and it is formed by these ingredients being pressed down and compacted for millions of years. Shale can be black, gray, reddish, or even yellow, depending on slight variations of the mineral content. You can easily break shale with a soft tap from a hammer.

When shale is buried under more sediment for a longer time, it creates slate. Slate can also form from volcanic activity. You'll find slate is harder and stronger than shale, and, like shale, slate can vary in color (dark red, gray, and green) or have streaks of color, although most slate sold for aquariums is gray. Slate is the most common rock offered for aquariums, usually sold as flat slabs ranging from a few centimeters long to over 12in (30cm).

Both slate and shale are inert and won't alter the pH or water hardness of your terrarium.

OTHER SUITABLE ROCKS

Frodo stone
Red island lava rock
White panther stone
Kita mountain rock

ABOVE Lava rocks are highly porous and a great choice for growing plants on.

DRAGON STONE

A popular and intriguing hardscape material, dragon stone is a distinctive-looking rock that boasts a rugged and weathered appearance with intricate textures, holes, and crevices. Its color palette typically includes rusty shades of brown and gray, and it often resembles the scales of a dragon—hence its name.

Dragon stone has a rough and porous texture that allows it to hold on to water, making it an ideal surface for creeping plants and mosses to climb up. It is considered inert and won't significantly alter the pH of the soil in a terrarium, but if used in an aquatic setup, dragon stone may release tannins into the water, especially when it is newly introduced. This can result in a slight yellowish tint, but this is generally harmless. Before you add this stone to any container, it is good practice to give it a rinse and scrub it thoroughly to remove any dust or debris.

SEIRYU STONE

This stone has become a hardscape favorite among aquascapers and terrarium makers. It's a type of limestone rock that originates in Japan and it holds striking characteristics. In Japanese, *seiryu* translates as "azure dragon," symbolizing the mythical creature believed to reside in blue or green waters.

Seiryu stone is typically a dark gray or blue-gray color and the intricate patterns and veins that run through its surface give it a rugged texture and an irregular shape that make it a visually striking addition to a terrarium. It's also a versatile hardscape material that can be used to craft visually appealing layouts, ranging from mountains and cliffs to captivating rock formations.

Beyond its aesthetics, Seiryu stone's high porosity is another beneficial feature. As it is highly porous, you should research which epiphytic plants can tolerate limestone rock before using it in a terrarium. It slightly raises the pH and hardness of aquarium water, so this must be taken into consideration if using acid-loving plants in a terrarium. As Seiryu is another dusty stone, you need to rinse it thoroughly to remove any loose debris before adding it to a terrarium.

ABOVE Dragon stone is so-named for its scalelike texture. It's a very popular hardscape for terrariums.

ABOVE Seiryu stone is a highly distinctive choice and particularly porous.

Wood, bark, and tree fibers

Wood needs to be prepared before use in a terrarium, and many kinds, especially softwood species, break down quickly and contain sap and resin that can upset the balance of the ecosystem. Fresh wood (meaning wood that has not had time to cure) will form mold quickly in a contained environment, so I highly recommend purchasing wood from a specialist supplier that has been treated specifically for use in a terrarium or aquarium. However, even this wood will go through a phase of forming mold. To avoid this, place your wood in a terrarium a few weeks before adding plants, and let the mold form and disappear first.

PETRIFIED WOOD

This is a fascinating wood because it's essentially a fossil. When trees are buried under sediment, they avoid decay due to lack of oxygen, and as groundwater slowly flows through the sediment it replaces the original wood with minerals such as silica, calcite, pyrite, and even opal. As the original wood is replaced by minerals, it transforms into what we know as petrified wood. Sometimes you can see patterns of the original tree bark or grain preserved in the rock. In the aquarium trade, you can find petrified wood that looks like a branch that's been cut into a log or shorter segment. The colors are usually gray/black or a reddish-orange. Petrified wood is inert, so it won't leach minerals. It's a truly fascinating and safe way to bring a piece of ancient history into your terrarium setup!

GRAPE WOOD

Derived from grapevines, *Vitis vinifera* is commonly known as grape wood or grapevine canes. It's another of my favored woods for a terrarium, due to its intricate grain pattern and naturally contorted appearance. Grape wood is produced during annual pruning, when excess canes and growth are trimmed to manage vine size and encourage healthy fruiting. When added to a terrarium, grape wood will go through a phase of forming mold, but this won't damage the plants. If you're using it in a vivarium (see page 17), put the wood through a mold cycle, as described on page 216.

MANZANITA WOOD

Manzanita wood is the common name for a group of shrubs and trees in the family *Arctostaphylos*, which are native to Northwest America. A versatile hardscape, it makes a beautiful centerpiece, or it can be broken down for an accent. Each piece is unique and comes in a variety of sizes, shapes, and textures. Some are twisted and gnarled, some are long and branchy. Manzanita wood decays slowly, which is a useful quality in a humid environment. This wood is smoothed and bleached in sunlight, which turns it from its usual reddish-brown into a light gray or white color. This transformation and its unusual shapes earn it the nickname "mountain driftwood."

SPIDER WOOD

Spider wood, also known as azalea root, is a popular type of driftwood used in terrarium building and aquascaping. It's one of the most common types of hardscape material for terrariums and gets its name from its unique appearance, which is said to resemble spider legs or webs. The wood is collected from the root system of azalea and rhododendron plants and comes in a wide variety of shapes and sizes—no two pieces are the same!

ABOVE Petrified wood pieces; the process of petrification takes up to 10,000 years.

It's a lightweight, versatile hardscape material that I prefer to break into smaller pieces to use as finishing accents. It looks particularly good when the curved pieces are placed around rocks and mosses. As a soft wood it is easily bent into shapes, but its pliable nature also means it rots over time and will eventually need replacing. It will also attract a layer of white mold when added to the terrarium, but this nearly always clears up without intervention.

CORK BARK

Another favorite of mine, cork bark has a natural resistance to rot and mold, and plenty of aesthetic appeal. The tiny holes, nooks, and crevices in cork bark provide perfect hiding spots for custodians.

Cork bark is harvested from *Quercus suber,* commonly called the cork oak tree, which is native to southwestern Europe and northwestern Africa. One of the unique aspects of the cork oak is that its bark can be harvested in large sections using special tools that don't harm the tree, and which even encourage the growth of new layers. The thick, outer-bark layer protects against environmental factors such as fire and extreme temperatures, but beneath it lies the valuable cork cambium, which consists of living cells, making cork oak a renewable source. Cork bark is typically removed every 9 to 12 years, and once harvested it is stacked and left to dry for several months. During this time, the bark undergoes a natural seasoning process, which makes it flexible and suitable for various uses. Once fully dried, the bark is ready to be transformed into different products. Lightweight, buoyant, and waterproof, cork also boasts excellent thermal and acoustic insulation qualities. It is available in many forms, including flat panels, thin and thick tubes, smaller pieces, and "caves" to be rested on the surface of the substrate or secured as a background.

TREE FERN FIBER

This fiber is used widely in terrariums, but is controversial in terms of its sustainability (see page 29). The panels can be used as a background or it can be broken into individual strands and used as accent pieces. I've only ever used tree fern fiber that's been directly harvested from nursery plants—I have never bought it.

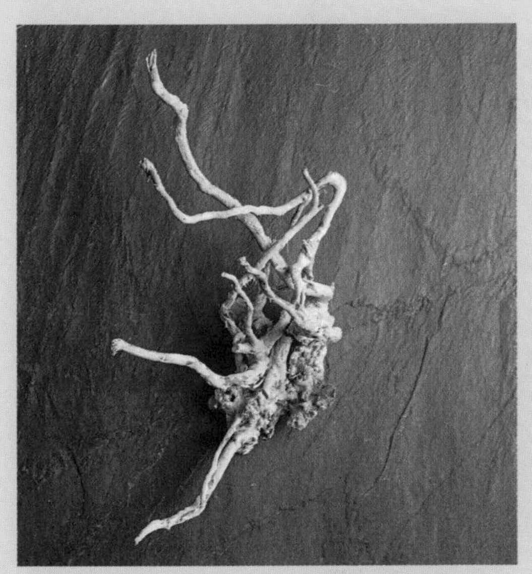

ABOVE Spider wood is one of the most widely used hardscape materials, which resembles spider legs.

ABOVE Cork bark; a multifunctional hardscape material, and a personal favorite.

MOPANI WOOD

This driftwood comes from the Mopane tree, *Colophospermum mopane*, which is native to the northern regions in southern Africa. Its highly unique appearance makes it one of the best types of hardscape materials available. It's characterized by its distinctive twisted and gnarled appearance and beautiful contrasting colors within the wood's grains. It's an incredibly tough species and is perfect for a terrarium as it will take a very long time to decompose! As it is so tough, it is not easy to break and you must use a saw if you want to reduce its size, but I much prefer to use it as a larger centerpiece anyway, rather than an accent. Preparation is simple: I just give it a once-over with a brush to remove any dust and debris then soak it in water for a day or two to remove any last bits of dust.

MALAYSIAN DRIFTWOOD/BLACKWOOD

Native to the forests of Malaysia, as the common name suggests, Malaysian blackwood (*Diospyros ebonasea*) is a unique, rare, and highly sought-after wood (and therefore expensive). It's easily identifiable due to its coloring, which sets it apart from other trees. It exhibits a dark brown color with streaks of black flowing throughout the wood, and some variations may appear so dark that they almost look entirely black.

Being an evergreen tree, it's considered one of the hardest softwoods that's used in commercial applications. *Ebonasea* wood tends to be less brittle than ebony woods found in other regions, making it highly regarded among vivarium enthusiasts. It's a heavy wood, and one that serves best as a centerpiece with epiphytes and mosses growing on it. I find it is best to soak Malaysian blackwood for a few days before adding it to your terrarium, to allow it to absorb water.

ABOVE Mopani wood is very heavy, so be careful with it as it can fall and break glass.

Detailing roots

Detailing roots can be bought from aquarium stores and do a great job of adding a finishing touch to a terrarium. I like using them with *Ficus* bonsai (see page 113), as they create the illusion of aerial roots that are present in mature trees in nature. They often go through a period of intense mold within the first few weeks, but this always passes with no issue. As the detailing roots are thin, they decompose faster in a terrarium than larger items and will need replacing.

Artificial hardscape

If you're not aiming for a completely natural look to your terrarium, you can use manufactured materials as background or stand-alone hardscape features. These can be linked up to a drip wall and are especially good for use in terrariums with misters, because they are often porous and very water-retentive, which means plants and mosses will easily grow on them. Generally speaking, I use natural materials because I prefer their appearance, but of course artificial materials will be covered by plants and mosses as they grow, giving a more natural look. These materials command a high price point, but they also work very well and are becoming increasingly popular among terrarium hobbyists.

EPIWEB

EpiWeb is made from a recycled polyethylene material that was designed to be a replacement for tree fern fibers (see pages 29 and 97). The material provides a lightweight and porous structure that retains moisture well and allows air circulation, creating a habitat for epiphytic plants and mosses to grow on and directly root into. EpiWeb comes in panels and is often used as a background material on the vertical surfaces of terrariums or vivariums. It can be easily shaped and cut to fit various designs, making it a versatile hardscape choice, and because it is inorganic it will not break down or degrade over time.

HYGROLON

Hygrolon is made from a blend of hydrophilic synthetic fibers—which means they effectively absorb and retain moisture. This product can hold up to 280 percent of its own weight in water and wicks up water vertically to a height of 24in (60cm). Thus it provides a consistent water source, mimicking the branches on which epiphytic plants grow in nature. Water that comes in contact with the outer layers of Hygrolon is spread over the entire surface, similar to how organic materials like cork bark, tree fern fibers, and sphagnum moss function, so it makes an excellent substitute for these natural materials.

Hygrolon is also very thin and can be used as a lining or background material on the vertical surfaces of terrariums or vivariums. It can be easily shaped and attached to various enclosure designs, adding a natural-looking and visually appealing support for plants.

Other interesting materials

You can really broaden your hardscaping choices and use items such as crystals, precious rocks, or even jewelry. Research whatever you want to include before use, though. I once accidentally added a piece of malachite to a terrarium, which is a potentially dangerous thing to do—with over 50 percent copper content, it is pretty toxic! There is great scope to be creative here, so feel free to mix up the use of interesting stones and crystals with natural elements such as branches, rocks, and plants.

ROTTING WOOD

Speaking with my good friend Adam from Micro Exotics, he told me that one of the keys to keeping colonies of isopods and millipedes healthy is to add ample amounts of rotting wood to the terrarium. I love the way he explained it to me, too: "You know when you're walking in the woods and you step on a log and it totally disintegrates under your foot? That stuff is perfect for them."

Since hearing this, I've collected small amounts of rotting wood and used that in bioactive terrariums I've made recently, and the custodians love it! Just be sure that you have permission from the landowner when collecting wood or you could get into trouble.

SEEDPODS AND CONES

The seedpods and cones from certain trees and shrubs make fantastic accent elements for a terrarium. They also benefit the ecosystem as they break down and are consumed by the custodians. I've noticed that my *Armadillidium* isopods love lotus heads and beech masts! Some options I love are alder cones, lotus heads/pods, coconut curls, trident pods, sororoca heads, mahogany pods, acorns, and beech masts.

Materials not to use in a terrarium

PRESERVED MOSS

***Leucobryum glaucum* is the most common type of moss used in moss preservation. This moss is actually dead, and the color in the leaves is retained using a chemical process. If you put preserved moss into a humid environment, mold will quickly develop on it and reduce it to mush. *Leucobryum glaucum* is primarily used in moss wall hangings or terrariums that have no live plant life. It seems strange to me to collect moss and soak it in chemicals that kill it (and any microfauna inside of it), so we can hang it on our walls.**

REINDEER MOSS

Not a moss at all, this is actually a lichen *(Cladonia rangiferina* and *Cladonia portentosa)*. It grows in acidic areas such as pine forests, montane heaths, and moors, and is a valuable food source for caribou, hence the name. It does not work in a terrarium because it prefers colder, aerated conditions.

LEAF LITTER

Leaf litter—the dead plant material that has fallen and dried from plants and trees—is an essential element in a bioactive terrarium, as it provides hiding places for custodians, as well as a food source for them. Even in a terrarium without custodians, leaf litter is a beneficial addition as it rots down over time, adding body and structure to the substrate. It's important to note that although it is great for insects, it won't do a good job of feeding plants, as it is not high in nutrients.

It is not often that I collect leaf litter, as it is available to buy from specialists, and the added bonus is that they offer a wider range of leaves! Jackfruit leaves, dried banana leaves, and pepper leaves are great additions and can be bought inexpensively. You can collect your own leaf litter if you'd like, but I would recommend allowing it to dry naturally, which will allow the "residents" to move on. It's important that it dries fully and that you inspect it before adding to your terrarium, just to make sure no one is in there. For example, grain mites will compete for food with springtails and can wipe out a colony quickly, while slugs and snails reside in leaf litter piles and will happily eat away at your lovely plants.

If you don't want to go through the hassle of sterilizing leaf litter, you can just allow it to dry naturally, which will allow the "residents" to move on. It's important that it dries fully and that you inspect it before adding it to your terrarium, just to make sure no one is in there.

While all leaves will work as leaf litter, I prefer to avoid using thick glossy leaves such as laurels, and spiky holly leaves (for obvious reasons). Oak, beech, hornbeam, and maples are all great native species whose fallen leaves make fantastic leaf litter; custodians will eat through them and often leave the leaf skeleton behind, which looks amazing in the terrarium!

RIGHT A great multipurpose material, fallen leaves make the perfect food and habitat for custodians, and are a natural soil additive.

How to create a background

MATERIALS

- Microfibre cloth
- Sandpaper or a sandpaper block for scratching the glass surface
- Black expanding foam
- Cork bark
- Crafting blade
- Vacuum cleaner
- Silicone
- Paintbrush
- Gloves (optional)
- Coir or sphagnum moss

This technique is usually only used on larger builds or vivariums. The expanding foam secures branches, rocks, and background panels in place, and once it has cured, it can be carved to create a naturalistic shape. It is then covered in silicone, to which sphagnum moss or a dried substrate material such as coir is added, to hide the foam. When I used to watch online terrarium-building guides, seeing people use expanding foam and silicone would put me off undertaking those projects, as it looked too complicated. But when I eventually tried it, I realized how simple it is to do! Please do this in a very well-ventilated area—or better yet, outside in an greenhouse or shed, as the silicone will smell while it is curing. Leave it for at least three weeks before adding any plants and custodians.

CONTINUED ⟶

1. Clean the inside of your terrarium to ensure no debris will sit between the foam and the glass. Once clean, use the sandpaper to lightly scratch the inside of the glass side where you want to affix your background. This will help the expanding foam adhere to it.

2. Carefully apply a thin layer of expanding foam over the areas you want to cover. Use the foam sparingly, as it really does expand a lot! Leave the foam for an hour to cure slightly—it should feel tacky to touch but still be soft underneath. Position the cork bark, pressing each piece firmly in place. Leave the foam for at least 24 hours to cure completely.

3. Once the foam has cured, take a crafting blade and carve off the shiny top layer of any remaining exposed foam. This will reveal a porous underside that the silicone will adhere to better. Continue to carve the foam until you are happy with the shape and texture of the surface, then clear away any foam debris with a vacuum. Be thorough, as any debris left on the expanding foam will get in the way when applying the silicone.

4. Apply a layer of clear silicone to the porous surface of the expanding foam and spread it out using a paintbrush or your finger. (If using your finger, use gloves, to prevent getting the silicone on your skin.) While it's still wet and before it has time to cure, apply a layer of dry coir or sphagnum moss to the silicone and press it in firmly so it sticks. Repeat this until the expanding foam is covered.

5. Use a paintbrush to brush away any loose bits of coir from the background. Examine the surface of the background. If any remaining areas of black foam show through, apply additional silicone and more coir.

6. Vacuum up any remaining debris and let the background stand for at least three weeks with the door open or lid fully removed, to allow the chemical smell to dissipate before adding any plants. By the time you plant it, there should be no chemical smell at all.

Fallen branch terrarium

MATERIALS

Cork bark tube with a hollow center
Bonsai concave cutters, or similar
Sphagnum moss
Dusk Moss Mix (optional)
Glass bowl
Pressurized spray bottle filled with distilled or deionized water
Wide fish tank container with an acrylic lid
Terrarium substrate
Tree fern fibers (optional)
Leaf litter (if using isopods)
Rotting wood (if using isopods)
Cuttlefish bone (if using isopods)

PLANTS

Begonia dodsonii
Begonia vankerckhovenii
Taxiphyllum sp.
Vesicularia montagnei
Selaginella uncinata
Elaphoglossum peltatum
Cochlidium serrulatum
Marcgravia sp. 'Mini Limon'
Pellionia repens
Solanum sp. 'Ecuador'
Dossinia marmorata

CUSTODIANS

Armadillidium isopods (optional)
Springtails (optional)

This terrarium project was inspired by the incredible creator and treasure trove of knowledge, Matthew Schwartz, founder of Another World Terraria, who hung a branch planted with epiphytic plants in a terrarium. This made me think about what would happen if a branch fell in the rainforest. So I decided to replicate this in a smaller scale, and cover it with epiphytic plants and mosses.

I chose a wide fish tank with a custom acrylic lid, as this gives the best aesthetic. I've simplified Matthew's project here; rather than attaching the branch to each side of the terrarium and hovering in midair, I rested it on the substrate, from one corner to another. Placing the branch at an angle allows more space for plants and mosses to grow.

CONTINUED ⟶

For the branch, I used a hollow cork bark tube filled with sphagnum moss. The moss gives the plants a medium to grow into and retains moisture, helping the terrarium to stay humid. I added Dusk Moss Mix to the open areas of sphagnum moss, which is easily made up from a packet and in a few months will grow into a dense carpet.

For the planting, I chose quite a lengthy and expensive plant list, but just use a selection of plants and aquatic mosses that are available to you. I added a mixture of *Taxiphyllum* sp. and *Vesicularia montagnei* mosses, as I wanted variety on the branch. Then I used small ferns *Elaphoglossum peltatum* and *Cochlidium serrulatum*, and the beautiful *Begonia dodsonii*. At the base I added a *Marcgravia* 'Mini Limon,' which will climb over the branch, and a *Begonia vankerckhovenii*, which produces beautiful yellow flowers. This wouldn't be a forest-floor terrarium without some *Selaginella uncinata*, to complete the planting.

Finally, to add a bioactive element, I used some *Armadillidium* isopods and springtails. These are forest-floor animals that will feed on the leaf litter and any decaying matter in the terrarium. I also added a healthy amount of rotting wood behind the branch, along with a piece of cuttlefish bone to provide them with a source of calcium. To find out more about bioactive terrariums, see Chapter 4.

1. Start by widening the opening along almost the whole length of the cork bark using the bonsai cutters. Fill the opening with sphagnum moss (keep some for later), but don't press it in too tightly. It should be airy and not too compact.

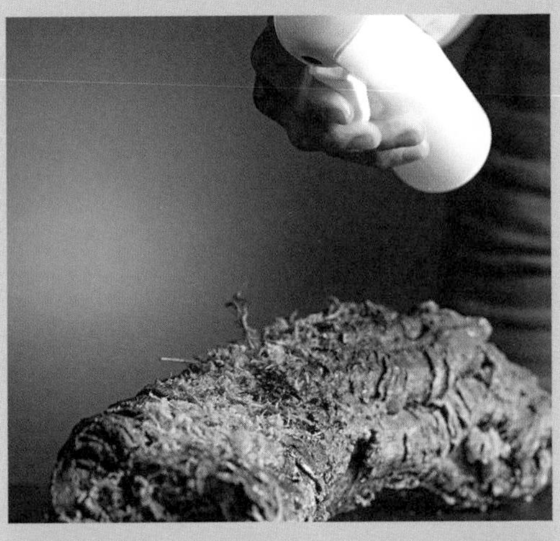

2. Spray the sphagnum moss inside the cork bark tube until it is just damp.

3. If you are using Dusk Moss Mix, make it up in a bowl following the packet instructions, covering it with water until it has a thin yogurt consistency. Set aside for as long as is stated on the packet—usually overnight.

4. Add your plants to the moss in an aesthetically pleasing way—I added the *Elaphoglossum peltatum*, *Cochlidium serrulatum*, *Begonia dodsonii*, and *Solanum* sp. 'Ecuador.'

4. Spread some of the soaked Dusk Moss Mix, if using, between the plants on the branch and on some of the openings on the cork bark. Cut the *Vesicularia montagnei* moss into small pieces and add directly to the branch.

5. Place the container on a sturdy surface and arrange the substrate on a gradient so that it is higher at the back and lower at the front. Top-dress with the tree fern fibers, if using, then spray the substrate with water until it is evenly damp.

6. Carefully add the planted cork bark to the terrarium, with one end in the corner where the substrate is lowest, and the other end in the corner where it is highest. It will span diagonally, from corner to corner.

8. Fill the space under the cork bark with sphagnum moss and add the terrestrial plants by making a small hole in the moss and positioning the plant inside.

9. Add the isopods and springtails, if you're using them, and healthy amounts of leaf litter for them to feed on. I prefer to add the majority of this leaf material behind the branch so it's out of sight, but a little at the front can also look nice.

10. Drill some small air holes in the lid. Place the lid on top of the container and situate it somewhere bright but out of direct sun (minimum of 200fc of light).

Bonsai in a terrarium

MATERIALS

Glass container big enough to hold your tree
Substrate—use the simple or premium mix on page 31, with the addition of 1 part composted bark
Pressurized spray bottle filled with distilled or deionized water
Fork for teasing out the *Ficus* roots
Petrified wood pieces or a hardscape of your choice
Long-handled scissors
Long-handled tweezers
Detailing roots (see page 98)

PLANTS

Ficus microcarpa 'Ginseng'
Leucobryum glaucum (moss)
Ficus punctata

The vast majority of trees used for bonsai are not suitable for use in a terrarium, as many temperate species require a dormant spell of temperatures below 41°F (5°C). Even tropical species that don't require this dormant spell can struggle in the confines of a terrarium. However, some *Ficus* sp. are an exception: *Ficus microcarpa* 'Ginseng' being the best choice as it's versatile and tough.

No tree in the world is going to like being in a stagnant environment, so choosing a suitable container with adequate airflow is important for this project. I've used a container made from sturdy recycled greenhouse glass. Using the right substrate for a tree is also important—I advise using a water-retentive medium that's predominantly made from fine-grained moler clay, akadama, lava rock, worm castings, sphagnum moss, and composted bark.

The substrate mix is airy, so ensure it doesn't fully dry out when it is in the terrarium. This can be especially tricky in a container that has no drainage holes! Lightly spraying the top of the substrate every now and again, when it starts to dry out, is the key to success with this type of terrarium, but remember to observe all of the soil and be extra careful not to overwater (see page 34).

CONTINUED ⟶

1. Place the glass container on a sturdy surface and arrange the substrate on a gradient so that it is higher at the back and lower at the front. Spray the substrate all over with water.

2. Remove the *Ficus microcarpa* from its pot and, using the fork, gently remove as much of the coir that the plant came in from around the roots as possible. Then trim away any thick anchor roots or any overly long roots (ensure you leave at least half of the root mass intact). This encourages finer feeder roots to form.

3. Make a hole off-center and higher up in the substrate, and place the tree into the hole. Firm it in place around the roots.

4. Add the petrified wood pieces into the desired positions. I've placed them around the base of the tree.

5. Trim away the beige parts from the moss using scissors (don't throw them away—see page 211) and position the green moss pieces around the front of the terrarium using long-handled tweezers.

6. Take some *Ficus punctata* cuttings and place them around the base of the tree, or wherever you choose.

7. Finish off the planting by adding some detailing roots. I find it looks best to anchor it from the top of the tree then place the other end into the substrate. This mimics the aerial roots formed in nature. Give the terrarium a final spray with water.

8. Place the lid on the container (opposite) and situate the terrarium somewhere bright where it'll get a minimum of 400fc of light.

Bonsai care

Like most bonsai, the tree will need to be removed from the container once every few years to have its roots pruned. This may seem barbaric but it's important for the health of the tree, and if this doesn't happen the *Ficus* will become rootbound and stop taking in water, eventually leading to its death.

As a general rule, repotting or root pruning is done in early spring, when the trees are in dormancy but just before they push out a new set of leaves. Tropical species are a little different and can be repotted throughout the year, but I still prefer to do it in early spring, as the weather is warming.

4

Inside the bubble

Keeping custodians

Not everyone wants to add "bugs" to their terrariums, but I like to do this because, for me, they add a layer of interest to this already fascinating hobby. They are not an essential requirement for a successful terrarium, but including them makes the terrarium bioactive, which can be beneficial to the terrarium ecosystem.

Detritivores like woodlice and springtails work at keeping the environment clean by eating dead plant material and waste, and reintroducing elements like carbon, nitrogen, phosphorus, calcium, and potassium into the soil. They do this by breaking down dead plant matter and releasing trapped nutrients from the plant tissues, making them available for plants to use for growth. It's fascinating how an abundance of detritivores in the soil allows the ecosystem to efficiently recycle nutrients, creating a sustainable and healthy environment.

Additionally, custodians help to improve the aerobic nature of the soil by creating tunnels and channels as they burrow and move through the substrate. This allows air and water to reach deeper into the soil, which is helpful for the plants' roots.

Some detritivorous animals, such as springtails, also eat harmful microorganisms including mold and fungi. By controlling the population of these harmful microorganisms, detritivores contribute to a healthier and more stable environment in the terrarium.

Isopods

Isopods, roly polys, woodlice, pill bugs, potato bugs; there is an exhaustive list of names for these amazing creatures and there are many, many different kinds, not all of which are suitable for use in a terrarium. While they may look like insects, isopods are in fact crustaceans and are closely related to crabs and lobsters! They have 14 legs and an outer shell called an exoskeleton, which is replaced with a new shell after molting.

There are many species of isopods, each with differing needs, so it's vital to undertake thorough research before adding them to your terrariums. As a general rule, all isopods need a source of protein and calcium, along with good amounts of leaf litter and rotting wood, to provide food but also to mimic their natural habitats and provide hiding places. (To protect themselves from predators, some woodlice can conglobate! This is when they roll into an impenetrable ball, much like an armadillo—the species *Armadillidium* is named after the armadillo!)

Isopod care

It's really fun to watch isopods go about their business inside the ecosystem, but there are some important considerations you must take into account before adding them.

Don't seal them in Custodians need air, so fully sealing containers with detritivorous animals inside won't work. Plants produce oxygen but the plants inside the terrarium don't produce nearly enough to sustain a healthy colony of detritivores.

Make them feel at home It is very important to mimic their natural habitat. For example, rotting wood is the main food source for millipedes, so adding plenty is vital to their survival, while isopods reside in and feed on leaf litter, so make sure this is available at all times. Springtails thrive in high humidity and won't tolerate drying out. Whichever species you include, please research them carefully.

Keep them apart It's best to keep isopod species separate; mixing them will always result in competition for food and space, with the end result being the demise of one colony.

Source custodians responsibly Do not take random isopods, snails, or millipedes from gardens or parks. Outdoors, they are in their natural environment and it's never certain if they will survive when transferred to a terrarium. The species I've listed in this section are tried and tested in terrariums and I always buy them captive-bred. Please purchase from a reputable seller and do not take any animals from the wild.

Isopods, like springtails, consume decaying organic matter such as dead leaves, plant debris, and wood. As they feed, isopods aid decomposition, converting the complex organic compounds into simpler forms and releasing essential nutrients back into the soil, ready for the plants to use. Interestingly, woodlice don't pee, instead, they expel toxins from their bodies by releasing ammonia through their shells. Isopods need calcium, but avoid adding sources of calcium such as crushed limestone to the substrate, because many terrarium plants dislike an alkaline substrate and it will have an adverse effect. Instead, I'd advise you to break off a piece of cuttlefish bone to leave in the terrarium, or add a supplement such as calcium powder during feeds.

While there are many species I could mention, on the following page I've listed my five favorite isopods for terrariums.

ARMADILLIDIUM KLUGII 'DUBROVNIK'

One of the easiest species to keep, *Armadillidium klugii* 'Dubrovnik' grows to a maximum size of ¾in (18mm) and has a native range from Croatia to Montenegro, which leads to a mixture of names from *Armadillidium klugii dubrovnik* to *Armadillidium klugii* 'Montenegro.' They have distinctive white, red, or yellow spotting on their backs, which gives them their common name, the clown isopod. They're fairly slow and can often be seen roaming around the terrarium, which I think is a great feature because fast-moving species often hide at the first sign of light or vibration. While this species doesn't tend to nibble on plants too much, it's important you provide it with a steady food source. It's a great beginner species.

CUBARIS MURINA

These small creatures grow no larger than ⅓in (10mm) in size. *Cubaris murina* are native to tropical and subtropical regions. They prefer habitats with high humidity as they are sensitive to drying out, making them perfect for use in a terrarium. They're unremarkable in appearance, usually a gray colour, but don't let this dissuade you from using them, as they're one of the most efficient woodlice species to use in a terrarium. They breed (and move) quickly and are one of the cheaper species to buy.

CUBARIS SP. 'RUBBER DUCKY'

This is the most expensive species on this list, with five being priced at around £100! They're so expensive because of the difficulties in breeding them, not to mention their distinctive appearance. A newly discovered species, *ubaris* species 'Rubber Ducky' has similarly shy behaviors to *Cubaris murina* and will spend much of its time hiding. Preferring a temperature range of 68–80°F (20–27°C) and a higher humidity, this species is perfect for a terrarium. They like to burrow deeper in a terrarium, so ensure there is at least 4–6in (10–15cm) of substrate.

PORCELLIO LAEVIS

Also known as dairy cows, this is a highly active species of woodlice that grows to a maximum size of (¾in) 20mm. Native to certain areas in Europe but found all over the globe, this species is usually lurking underneath fallen logs and in leaf litter. *Porcellio laevis* is highly effective at processing decaying plant matter but it's an opportunistic feeder, which makes it problematic for use in a terrarium, as it can quickly decimate your plants! To protect your plants, these creatures need lots of supplementary feeding. Still, they are a fascinating species to keep and they breed quickly in ideal conditions—just keep any rare plants away from them!

TRICHORHINA TOMENTOSA

Otherwise known as dwarf white isopods, these creatures hold a special place in my heart as one of the tiniest and most accessible isopod species out there. Growing to only 3–4mm, they take the crown as the smallest isopods I've encountered, but don't let their size fool you—even though they're tiny, they breed fast and are suitable for terrariums of all sizes. However, they are prolific, so they are best on their own in a terrarium and not with other species, as they will simply outcompete them!

They're not the most active species and will spend much of their time hiding in the substrate, but they are incredibly efficient when it comes to devouring decaying matter, and their energy is unmatched.

Dwarf whites are rapid breeders and are parthenogenetic, meaning they reproduce asexually. This species requires high humidity to thrive and the creatures are less fussy about ventilation needs compared to other isopods, but it's still a good idea to ensure some airflow. Give them lots of humidity, hiding places, leaf litter, and rotting wood and you'll have hundreds of these tiny isopods in your ecosystems. I've found that this species is particularly fond of a protein source and will quickly consume any fish food they're given!

CLOCKWISE FROM TOP RIGHT Isopods bring extra life and interest to terrariums; some of my favorites are: *Armadillidium klughii* 'Dubrovnik,' *Porcello laevis* (dairy cow), and *Cubaris* sp. 'Rubber Ducky.'

Millipedes

Millipedes are fascinating creatures best known for their numerous legs, with most species having anywhere between 30 and 400, although the number varies depending on the species. Most have long, cylindrical bodies that are divided into segments that typically bear two pairs of legs. Like isopods, millipedes have an exoskeleton that is usually dark brown or black, but some species have colorful markings or patterns.

Millipedes are slow in their movements, with their legs working in a wavelike motion as they crawl. When threatened, they can curl their bodies into a tight spiral, forming a protective ball that shields their softer, vulnerable underside and exposes their harder, armored exoskeleton. Some species can also release a secretion that can be toxic to predators—and are even known to secrete cyanide! Don't handle any kind of millipede; if you absolutely have to do so, wear gloves.

Millipedes, like most of my favorite custodians, prefer moist environments and are often seen scurrying around in leaf litter, decaying wood, and other organic matter on the forest floor. Some species are also known to burrow in the soil, which, within a terrarium, is useful for keeping it aerobic. It's important that millipedes are not pulled off surfaces or disturbed when molting, as this can prove fatal to them.

Millipedes aren't as effective at breaking down organic matter and are more specialized in their diet, preferring softer decaying organic matter and occasionally feeding on live plant roots when food is scarce. It's essential that millipedes have large amounts of rotting wood to feed on as this is their main food source.

OPPOSITE *Desmoxytes planata* on a piece of rotting wood, which is its main food source.

DESMOXYTES PLANATA

These pink dragon millipedes are highly recognizable, with flattened bodies and a prominent ridge running along the length of their backs. Reaching lengths of 1in (30mm), they have a distinctive coloration, with their exoskeleton ranging from dark brown to black, with bright pink spurs. This fascinating animal was discovered in 2008 and is native to the Andaman Islands of India, but it has a vast distribution over tropical regions. It is a rare species, and its price has shot up in the past year. *Desmoxytes planata* are fragile and this species is known to secrete cyanide from specialized glands, so please don't handle them without gloves on!

Desmoxytes planata need high humidity and good airflow to thrive. In nature, they're found in limestone caves, which signifies that they prefer an alkaline environment. It's essential that these millipedes have an abundance of rotting wood in the terrarium, along with a supplementary supply of protein and calcium. I've found that oak (*Quercus robur*) is their preferred wood of choice, which I shred into tiny pieces to help it decay faster. Often, horticultural bark comes from pine trees, which is on the acidic side, so I would avoid adding it to any terrariums that house this species of millipede.

Once they're settled, dragon millipedes tend to breed prolifically; their babies are minute and can be spotted climbing on the glass of the terrarium.

ANADENOBOLUS MONILICORNIS

Bumblebee millipedes have a cylindrical body that's divided into numerous segments, each bearing two pairs of legs. Their exoskeleton features unique black and yellow bands that resemble the color pattern of bumblebees, hence their common name. Their body length can vary, but they typically reach around 1½in (40–50mm) when fully grown. They have become popular in the terrarium hobby due to their unique appearance and simple care requirements. Unlike *Desmoxytes planata*, *Anadenobolus monilicornis* spend most of their time in the top few inches of the substrate, so you can use a thinner layer of your growing medium if you introduce these creatures. You should provide ample amounts of leaf litter and rotting wood for this species to help them thrive.

Snails

The only kind of snail I've ever added to a terrarium is *Oxychilus alliarius*, a tiny species of terrestrial snail which is commonly known as the garlic snail due to the distinctive garliclike odor it emits when disturbed. (*Alliarius* comes from the Latin word *allium*, for garlic.) As far as I'm aware, there is no place to buy these snails so if you want some, you'll have to rely on a few hitchhikers coming in on a potted plant! They are minute, getting no bigger than 6–8mm.

They're native to regions in Europe and can be found in many kinds of habitats, including forests, grasslands, and even your gardens. They're primarily nocturnal creatures, so they are more active during the night and seek shelter during the day to avoid drying out. They are generally slow-moving and feed on decaying plant matter and detritus—unlike common garden snails, they largely ignore your healthy plant foliage. I have noticed that when I feed the custodians in my bioactive terrarium, the snails always go for the fish food and I've seen them eat dead crickets, which leads me to believe they are fond of protein sources! It's also important to make sure there is an ample supply of calcium, so their shells remain healthy. You can provide this by adding cuttlefish bone.

Oxychilus alliarius is hermaphroditic, meaning each individual has both male and female reproductive organs. However, they still require mating with another individual to exchange sperm and fertilize their eggs. After mating, the snails lay eggs in moist soil or leaf litter, and the eggs hatch into small juvenile snails. These little snails have a closely coiled, translucent shell and a dark blue body. While garlic snails will eat the leaves of certain plant species, I feel they're worthy of a mention.

LEFT *Oxychilus alliarius* emits a strong garlic smell when disturbed.

Worms

While on paper worms may seem like a helpful addition, they need a specific environment that the terrarium cannot provide. Worms commonly sold as fishing bait or used in worm bins are composting worms like *Eisenia fetida* or *Perionyx excavatus*, which reside in the upper few inches of soil. They're fed large amounts of organic matter, which they consume quickly, meaning they need to be fed frequently. In addition to this, worms need good amounts of airflow. While the odd rogue worm may find its way into a terrarium, most likely coming in as an egg in some worm castings, I highly advise you do not add worms to your terrariums. If you see any in there, I would recommend removing them.

Springtails

Think of the humble springtail as a janitor of the terrarium. These tiny hexapods are usually no bigger than a pinhead and are found in moist conditions that are rich in organic matter. Springtails are primary decomposers in the terrarium and feed on decaying organic matter, such as rotting plant material, fungi, and bacteria. It's processed in their digestive systems and the nutrients are released in a form that can be absorbed by plants and other organisms in the substrate. It's often said that springtails help aerate the substrate, too, but I'm sceptical about this; their size and modest numbers in the terrarium makes me think their impact in this sense is negligible.

By feeding on mold and decaying matter in a terrarium, springtails are effective at preventing small mold outbreaks becoming larger, but it's important to note that correct practice, such as using a high-quality substrate and providing adequate airflow, is more important in keeping mold at bay. Adding springtails to a terrarium covered in mold will not have any effect. They're only effective once an established colony is present. Springtails reproduce through a process called parthenogenesis, where females are able to produce offspring without mating. Establishing a colony, in the right conditions, is simple and fast; the life cycle of these springtails is fairly short, with it taking only a few weeks to see a complete generation.

Be careful when opening the lid to your container, because true to their name, springtails can jump many times their own body length. Don't worry, though, any escapees won't survive in your home.

The most common springtails used in terrariums are the tropical white springtail, *Folsomia candida*, and the orange springtail, *Bilobella braunerae*.

FOLSOMIA CANDIDA

This little springtail measures 1–2mm in length. While they are small, they're still visible inside the terrarium and can often be seen climbing on the sides of the container and over any decaying matter. They have a distinctive white or pale coloration, which has earned them the common name "white springtail." Their body is elongated and soft, with six legs and a pair of antennae, and under a macro lens they look stunning. This is the most popular species to use in a terrarium.

BILOBELLA BRAUNERAE

An unusual-looking species of springtail that looks like one of those puffy cheesy crisps. Only discovered in 1981, *Bilobella braunerae* is a jumpless form of springtail that reaches a maximum size of 4mm. As a decomposer, they're not as effective as *Folsomia candida*, but they provide a high level of interest in a terrarium, and under a macro lens their chubby bodies look very cute.

How to culture springtails

Springtails make a wonderful addition to your terrarium and are so easy to culture. All you need is one starter pack, which you can buy from most reptile stores, and in a few weeks you can expand the colony significantly.

I've used charcoal here, rather than a soil-based medium, as it makes removing the springtails far easier. It's vital you use clean, organic charcoal—not the stuff you use for a barbecue, as these often come with a catalyst to help them ignite. I prefer to use plastic deli cups when making cultures, as they're inexpensive and durable.

When starting a culture it's important not to overfeed, because as the food decomposes gases are released, which can kill off a colony quickly. So keep springtails happy by feeding little and often; I prefer using active yeast, but people often offer uncooked white rice instead. I think it's better to have lots of smaller cultures rather than one big one, so if something goes wrong in one colony, you don't lose them all.

MATERIALS

Organic, catalyst-free lumpwood charcoal
Large bowl of water
Plastic bag for breaking the charcoal
Hammer
Sturdy plastic deli cup with a lid
Needle
Pressurized spray filled with distilled or deionised water
Sprinkling of active yeast

CUSTODIANS

Existing springtail culture

1. Place a few charcoal pieces into the large bowl of water and leave to soak for 1 hour.
2. Drain, then place the wet charcoal into the plastic bag and, using a hammer, carefully break the pieces into 1–2in (2.5–5cm) pieces. Do this outside as it is dusty even when damp. Rinse the charcoal pieces so no dust is left.
3. Remove the lid from the deli cup and use a hot needle to pierce tiny holes. Place charcoal inside to fill three-quarters of the tub. Add water to fill the bottom fifth of the container.
4. Carefully add a spoonful of your springtails from the existing culture into the deli cup. If they're in a soil-based medium you will have to add a little soil to the new container to aid the transfer; if they're in a charcoal medium, simply add a few pieces of charcoal.
5. Add a small sprinkle of active yeast to the new container, replace the lid, and place somewhere dark. It's a good idea to open the lid once a day to allow fresh air inside.

Bioactive terrarium

MATERIALS

Mesh
Permanent marker
Scissors
18in (40cm) diameter fishbowl, and a lid
Expanded clay pebbles for drainage
Terrarium substrate
Pressurized spray bottle filled with distilled or deionized water
Large piece of dragon stone as a centerpiece
Long-handled tweezers
Leaf litter and shredded rotting wood
Acrylic lid

PLANTS

Nephrolepis ferns
Ficus thunbergii
Ficus punctata
Biophytum sensitivum

CUSTODIANS

Custodians of your choice

This is a variation on my famous bioactive terrarium (see page 18). The centerpiece is a large piece of dragon stone, which gives the creeping figs a surface to climb on and the custodians somewhere to hide. I've also added a preying mantis model from the LEGO® IDEAS Insect Collection, as it fits in so nicely with the theme of this terrarium. It's important to provide the custodians with a good amount of leaf litter, which I do last so I can cover any substrate that is showing. I've added *Desmoxytes planata* millipedes, *Folsomia candida* springtails, *Oxychilus alliarius* snails, and *Armadillidium nebula* woodlice here. Use whatever you like, just don't mix isopod species. For the planting, I've used cuttings of *Ficus punctata* and *Ficus thunbergii*, and the mini palmlike plant *Biophytum sensitivum*, which grows like a weed in Southeast Asia. I left out mosses, as hungry custodians like to munch on them.

CONTINUED ⟶

1. Place the mesh on a sturdy surface and draw around the circumference of the bowl with a permanent marker. Cut out the mesh circle and set to one side. Place the container, right way up, on the table and add a 4in (10cm) layer of expanded clay pebbles in the base.

2. Cover with the mesh, then add enough substrate until two-fifths of the terrarium is full, arranging it on a gradient so it's higher at the back, lower at the front. Spray with water until evenly damp.

3. Firm the dragon stone centerpiece into place. Be sure it's firmly in place, as you don't want it to rock or fall later on!

4. Remove the plants from their pots; position the *Nephrolepis* ferns at the back of the container, with the *Biophytum* to the side. Using long-handled tweezers, firm the *Ficus* cuttings around the base of the container but pointing toward the dragon stone, so they can climb on it. Gently spray all the plants with water.

5. Assemble the LEGO® set (if using) following the instructions in the box. Place the assembled LEGO insect in the container, toward the back and slightly off-center.

6. Drill some small air holes in the lid. Add a healthy amount of leaf litter and shredded rotting wood over the substrate. Carefully tip the custodians out of their container (from a low height!) into the terrarium, then secure the lid on the container.

Jumping spider terrarium

Definitely not one for arachnophobes! Jumping spiders are a fascinating group of arachnids, known for their unique hunting strategies, incredible jumping abilities, and often amusing appearance. I think these spiders are cute; they're also easy to keep and a good species for beginners.

Found worldwide, with thousands of known species, they are one of the largest, most varied families of spiders, but the best species for a terrarium is *Phidippus regius*. Jumping spiders have compact, sturdy bodies ranging from a few millimetres long to ¾in (2cm). Their eight eyes are arranged in a distinctive pattern on the front of their head, with the large pair of eyes—the anterior median eyes—providing excellent vision and playing a crucial role in their hunting behavior. Unlike other spiders that spin webs to catch prey, jumping spiders are agile hunters that use their silk mainly for safety lines and egg sacs. When hunting they move slowly and stealthily toward their prey, then use a sudden, precise leap to pounce on their target. Their hunting success is aided by their surprisingly high level of intelligence compared to other arachnids; they can learn from past experiences, recognize and remember specific individuals (you!), and adjust their behavior accordingly.

This is a simple enclosure but it's important not to densely plant, or the spider can struggle to find food. The container must have ventilation holes and be front-opening, as the spiders nest at the top of the container, and opening the lid will destroy this.

Jumping spiders won't survive in a normal terrarium. It's imperative that you research the care requirements of your specific jumping spider for this project so that you are comfortable with its needs.

MATERIALS

A front-opening container
Any terrarium substrate
Pressurized spray bottle filled with distilled or deionized water
Cork branch as hardscape
A piece of Seiryu stone
Fine lava rock 1–3mm in size, for top-dressing

PLANTS

Ficus microcarpa 'Ginseng'
Pilea cadierei

CUSTODIANS

Adult *Phidippus regius* spider
Small crickets or bottle flies

NOTE Jumping spiders are not considered dangerous to humans; their bites, if they occur, are harmless and may only cause mild irritation. However, as with any wildlife, it is best to appreciate them from a respectful distance and avoid handling them.

1. Add a 1½in (4cm) layer of substrate to the base of the container. Spray the substrate with water until it's evenly damp, then firm the cork branch and Seiryu stone into the substrate.
2. For the planting, remove the *Ficus microcarpa* 'Ginseng' from its pot and plant it toward the back, slightly off-center. Take cuttings from the *Pilea cadierei* and position on the opposite side of the container.
3. Carefully take the lid off the spider container and lower it into the terrarium. Don't force it out—patiently wait for it to come out on its own accord. Once the spider is out, remove the container and offer it a cricket to feed on.

NEXT PAGE *Phiddipus regius*, jumping spider (left); and *Hymenopus coronatus*, orchid mantis (right).

Praying mantis terrarium

Also known as mantids, praying mantises are incredible insects with distinct features, such as a slender body and a triangular head that can rotate up to 180 degrees. They display ferocious predatory behaviour, and as carnivores they primarily hunt and feed on other insects, using their front legs to grab and secure them. Praying mantises have evolved remarkable camouflage adaptations; some mimic the appearance of leaves, twigs, or flowers while they wait to ambush their unsuspecting prey.

I chose an orchid mantis (*Hymenopus coronatus*) here because of their pink and white coloration and relatively basic care needs. This species comes from the tropical forests of Malaysia, in a natural habitat of white and pink flowers in bushes and small trees. Thus the lobes on their legs mimic flower petals, to camouflage them as they lie in wait for pollinating insects attracted to the flowers. Mantises can change color in a matter of days, depending on environmental conditions such as humidity and light; this species needs a relatively high humidity of 60 to 80 percent and a steady temperature of 64 to 82°F (18 to 28°C).

MATERIALS

A mantis-specific container (it should have ventilation holes and a mesh lid—see box, right)
Sphagnum moss
Terrarium substrate
Pressurized spray bottle filled with distilled or deionized water
Tree fern panel to act as a background
Fine lava rock

PLANTS

Ficus microcarpa 'Ginseng'
Tillandsia stricta
Pilea cadierei

CUSTODIANS

An adult orchid mantis

Mantis care

All praying mantises need an enclosure at least three times the length of the insect in height, and twice its length in width, anything larger and it will struggle to find its prey. For an adult female that's at least 10in (24cm) high and 6in (16cm) wide. The container needs a mesh on top for the mantis to grip onto, to prevent it falling.

Mantises should feed every 3–4 days, but they are not the best hunters. If the mantis isn't catching prey, move it to a feeding container to eat, then return it to its home. A varied diet of houseflies and bottle flies is ideal, but add a moth or hoverfly if you find one. Avoid using crickets as they can attack the mantis. Mantises will gorge themselves if given the chance.

1. Place the container on a sturdy surface, then press a layer of sphagnum moss into the base of the container. Scatter over a 2in (5cm) layer of substrate and spray until it's evenly moist. Firm the tree fern panel into the substrate so it covers the entire background.
2. For the planting, take the *Ficus microcarpa* 'Ginseng' out of its pot and loosen the root ball. Trim the roots and plant into the substrate. Top-dress with the fine lava rock and place the *Tillandsia stricta* on top.
3. Take cuttings from the *Pilea* and place the cuttings at the front of the container.
4. Open the container the mantis came in, then carefully lower it into the completed terrarium and let the mantis come out in its own time. Then remove the container.

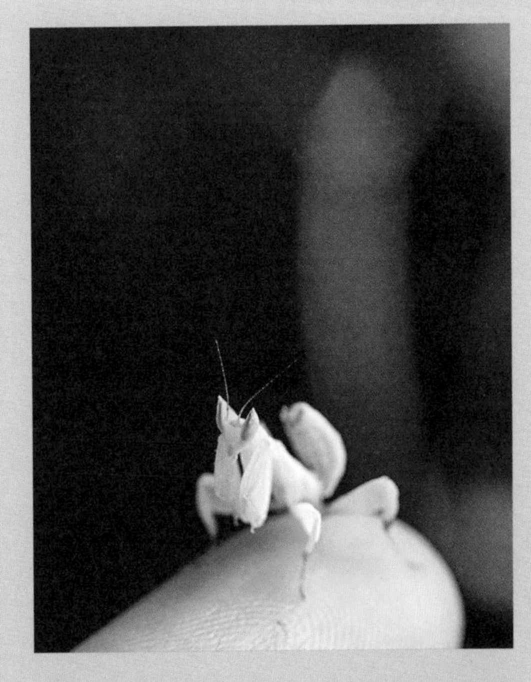

5

All about plants

Ferns

Ferns are some of the oldest plants on Earth, having been around for millions of years. They're a type of nonflowering plant that belongs to the group known as pteridophytes, which includes horsetails (*Equisetum)* and whisk ferns (*Psilotum*).

Pteridophytes are considered primitive vascular plants, which means they have specialized tissues for conducting water and nutrients throughout their structures. Unlike flowering plants (angiosperms), which produce seeds, pteridophytes reproduce through spores. These spores are typically produced in structures called sporangia, which are located on the undersides of fern fronds or within specialized cones in horsetails. The fronds, or leaves, of ferns are typically divided into smaller leaflets called pinnae. These leaflets may be further divided into smaller segments called pinnules. The fronds unfurl from a coiled structure known as a crosier.

Ferns come in a wide range of sizes, from small-leaved species to large tree ferns that can grow several meters in height. These plants can be found in various habitats, including forests, wetlands, and tropical rainforests. They tend to thrive in areas with high humidity and shade, making many species perfect for use in a terrarium. As a general rule, I make sure the ferns in my terrariums get a minimum of 150–200fc of light, but some species may need more or can tolerate less. These are some of my favorite ferns for a terrarium.

ACTINIOPTERIS AUSTRALIS

This is a delicate fern with palmlike leaves. I've found it to be difficult to keep, as it requires steady high temperatures (+ 68°F/20°C) and high humidity to flourish and grow to a maximum size of 6in (15–16cm). It thrives in lower light conditions and will quickly turn brown if exposed to too much light or if it is allowed to dry out.

ASPLENIUM AFF. FRAGRANS

This delicate little fern is native to South America and grows to a height of 6–8in (15–20cm). It thrives in a slightly lower light range of 100–150fc and requires high levels of moisture. It can grow epiphytically (on other plants, but without feeding on them) if the humidity is high enough, but I've had most success growing this beautiful fern in a substrate of straight sphagnum moss.

Asplenium sp. 'Zamora'

ASPLENIUM SP. 'ZAMORA'

A beautiful, tiny fern at just about 2½in (6–7cm) that is native to South America. It does well in bright indirect light of around 200fc, when grown in a layer of sphagnum moss and high humidity but with ample airflow.

COCHLIDIUM SERRULATUM

This charming tiny fern can be found in South America, mainland Africa, and Madagascar. Everything about this beautiful fern captures my attention: from its thin, serrated leaves to the way the spores cling to the underside of the foliage. This is a new addition to my collection but I'm excited to propagate it and try it in a variety of settings. It grows up to 2½–2¾in (6–7cm).

DORYOPTERIS CORDATA

An unusual tropical fern, *Doryopteris cordata* is well suited to cultivation in a terrarium. It forms a low-growing mound of leaves and featherlike fronds that extend 6in (15cm) above the plant, which is what gives it the common name of antenna fern. This fern requires high humidity, warm temperatures, a semishaded location away from intense light, good airflow, and moist, well-draining, fertile soil that should not be allowed to dry out.

ELAPHOGLOSSUM PELTATUM

A stunning, highly recognizable epiphytic fern from Ecuador that is simple to grow as long as it has high humidity and a minimum of 100fc of light. It's best grown mounted onto bark or a tree fern with the rhizome set slightly above a bed of sphagnum moss.

HEMIONITIS ARIFOLIA

This easy-to-grow terrarium plant is native to tropical Asia and has dark green, heart-shaped leaves at the end of long, wiry stems. It can grow happily as a houseplant but it requires high levels of humidity to thrive. An unusual feature of this fern is how it propagates: baby plants appear at the end of each stem toward the base of the leaves. Once roots form on these, the baby plants can be removed and planted on into a pot of substrate or straight into a terrarium.

Cochlidium serrulatum

Elaphoglossum peltatum

MICROGRAMMA VACCINIIFOLIA

Native to South America and the Caribbean, *Microgramma vacciniifolia* is an epiphytic beauty. It requires warmth and humidity to thrive, and in a terrarium it does best when mounted on bark or grown over a background. This is another new species in my personal collection, and I'm keen to see how it takes.

NEPHROLEPIS CORDIFOLIA 'DUFFII'

This is the first of three entries in this list from the *Nephrolepis* genus. This wonderful tender fern is native to Asia and Australia and has the common name of Boston fern, or the lemon button fern, due to the buttonlike leaflets along each frond and the pleasant aroma that is released when the leaves are handled. If allowed to, it will spread; it can reach a height and width of over 12in (30cm), but it can be pruned or divided to keep it small. It thrives in high levels of humidity.

NEPHROLEPIS EXALTATA 'FLUFFY RUFFLES'

I'm not sure who thinks up these names, but this one is kind of cute. *Nephrolepis exaltata* 'Fluffy Ruffles' goes by the common name of sword fern due to its elongated, fresh, green leaves. It reaches a size of 20in (50cm) if given free rein, which is far too large for most terrariums, but it is happy to be pruned and kept small.

NEPHROLEPIS EXALTATA 'MARISA'

My favorite of the *Nephrolepis* genus, *Nephrolepis exaltata* 'Marisa' is a delightful fern with tufted, lacelike fronds. It stays fairly compact and should be thinned out occasionally to stop the leaves in the center from dying. It does best in medium to high humidity.

Microgramma vacciniifolia

Nephrolepis cordifolia 'Duffii'

LEMMAPHYLLUM MICROPHYLLUM

Commonly known as the Japanese beard fern, this is a beautiful creeping fern from Southeast Asia. It has circular emerald leaves and can be found growing epiphytically on moist tree bark. It's an easy plant to care for in a terrarium, as it requires low to medium light, and it propagates easily via stem cuttings. It grows well terrestrially and epiphytically.

PELLAEA ROTUNDIFOLIA

Also known as the button fern because of its alternating circular leaves along the stem, this is a tender fern that's endemic to New Zealand. It is easy to care for in a terrarium as long as it has an airy, water-retentive substrate that isn't allowed to dry out.

PTERIS ENSIFORMIS 'EVERGEMIENSIS'

Commonly known as the silver lace fern, this is a beautiful variegated fern that's native to the UK. It requires slightly more light than many species mentioned here and does benefit from an east- or west-facing window. It grows large and takes up a lot of space in the container (not always a bad thing!), but can be kept in check with a little pruning.

Lemmaphyllum microphyllum

PTERIS QUADRIAURITA 'TRICOLOR'

When the fronds of this stunning fern unfurl they are a bright pinkish-red, but as time goes on they turn bronze before settling on green when mature. This fern doesn't like to dry out, but don't water it with hard tap water as it requires a slightly acidic substrate. Plants can grow quite large, and will need pruning to keep small.

PYRROSIA NUMMULARIFOLIA

The creeping button fern is not only one of my favorite ferns but one of my favorite plants. It's quite rare but it is so easy to care for, which is the best combination! Native to Indonesia, it has small, rounded leaves and can be easily grown epiphytically or terrestrially. It requires higher light levels than the other species on this list, otherwise the nodes between the leaves become larger—I recommend a minimum of 150fc. I find this fern grows well on a more water-retentive medium topped with sphagnum moss. If grown epiphytically it will need frequent misting. It propagates easily by stem cuttings.

Ficus

The *Ficus* genus is a diverse group of plants that belongs to the mulberry family (*Moraceae*), which is one of the largest groups of flowering plants (angiosperms), with over 800 recognized species. Ficus plants are native to tropical and subtropical regions around the world, including Asia, Africa, and North and South America.

They are known for their distinctive foliage, which can be glossy, leathery, or deeply lobed, and they exhibit a wide range of growth habits and sizes—from invasive creeping forms like *Ficus pumila*, to large trees such as *Ficus carica*, which are known for their edible fruits. The fig inflorescence is not a traditional flower but a complex structure that encloses many tiny flowers, which all contain seeds.

The *Ficus* that we use in terrariums are the small- to medium-leaved species, and there are some stunning varieties available. Be careful when handling these plants, as the milky latex that is contained in the leaves and stems can be an irritant and can cause a nasty rash. So keep the plants well away from your eyes and always wash your hands thoroughly after handling them.

FICUS SP. 'BORNEO'

Native to Indonesia, this beautiful trailing *Ficus* has stunning leaves, with new growth that's light green maturing to a deep green. As the plant matures it also develops a woody stem. It propagates easily via cuttings. It has a slow growth rate, making it an ideal choice for terrariums of all sizes, and does best in a water-retentive, airy medium where it will act as a ground-cover or a climbing plant, if given a surface to grow up.

Ficus sp. 'Borneo'

FICUS SP. 'BORNEO SMALL'

This has to be the quintessential terrarium plant. It's a smaller-leaved, even-slower–growing version of *Ficus punctata* (see below) that's native to Borneo. A striking plant that is suitable for use in the smallest of terrariums, it needs the same conditions and care as other *Ficus.*

FICUS PUMILA

This woody plant has a fast growth rate. It requires high levels of light to reach its maximum size of 3ft (1m), so the confines and lower light levels of a terrarium naturally keep it in check and significantly smaller, with only a little diligent pruning required. Native to East Asia, this plant has since spread into many warm places in the world, such as Australia, the southern states of the US, and the Mediterranean—where it can be seen taking over buildings. For use in a terrarium, I recommend taking rootless cuttings with at least three leaves on (see page 208) and either placing the stems directly into the soil or resting them (leaves facing up) on a layer of moist moss or substrate. The plants will root quickly in a humid environment. *Ficus pumila* 'Sunny' is a pretty, variegated version that has the same growth rate as *Ficus pumila*.

FICUS PUMILA VAR. MINIMA

A smaller-leaved, slower-growing version of *Ficus pumila* that's native to Colombia, *Ficus pumila* var. *minima* (also known as *Ficus colombia*) is a lowland rainforest plant with a creeping habit. It requires high levels of humidity (of at least 70 percent) and does well grown from rootless cuttings. I recommend taking cuttings with at least three leaves on (see page 206) and resting them on a bed of moist moss or a moisture-retentive substrate. This *Ficus* makes a wonderful ground-cover or background-covering terrarium plant.

Ficus sp. 'Borneo Small'

Ficus pumila

FICUS PUNCTATA

An easy addition to this list, the lance leaf fig is one of my favorite plants. Sometimes called *Ficus panama*, which is strange because it is native to Southeast Asia, this fig is a slow-growing, easy-to-care-for terrarium plant with dark green, oval-shaped leaves. It prefers a humidity range of over 70 percent and a moisture-retentive substrate that doesn't dry out. It grows well terrestrially and will form a dense mat if it has nothing to climb, but it will also grow up and over the surface of a tree, rock, or the background in a terrarium with ease. You can easily propagate this plant by taking cuttings of five leaves, then removing the bottom leaf and placing the stem into a moist substrate or a layer of sphagnum moss.

FICUS SAGITTATA

A terrestrial creeping vine that's native to Northeast India, the Andaman Islands, and Southern China, it has dark green leaves that are oval in shape and have a leathery texture. *Ficus sagittata* grows best under strong light and its large leaves mean it does well in medium to large terrariums. Don't let it dry out. It propagates well with rootless cuttings.

FICUS THUNBERGII

Also known as *Ficus quercifolia*, the oak leaf fig is one of the most popular terrarium plants—and for good reason! Native to China and East Asia, *Ficus thunbergii* has tiny leaves that are similar in shape to oak leaves. It forms a dense mat of foliage under good light, which acts as a stunning ground cover, or it will readily climb the backdrop of a terrarium. New leaves are a vibrant light green but get darker with age. It prefers substrate that is moisture-retentive and does not dry out. The leaves grow larger in lower light conditions, so I advise giving this plant a minimum of 200fc to ensure nice compact growth. This plant is unfussy if it has moderate to high humidity.

FICUS VILLOSA

This terrestrial climber is widespread over Asia—although it is critically endangered in some areas. *Ficus villosa* requires high humidity and a moist substrate to thrive. It is a larger-leaved species that does best when it is allowed to climb a background. All new growth is bronze in color and turns a darker green as it matures.

OPPOSITE *Ficus punctata*

Mosses

It's often said that mosses complete a terrarium—and I couldn't agree more. There's something infinitely satisfying about moss in itself, and watching it grow in a terrarium brings so much joy.

Long before I was interested in plants, I remember being fascinated by the vibrant green moss I saw in the Forest of Dean; I thought there was something incredibly enchanting about the way it grew over the forest floor. Moss will shrivel and dry out in periods of drought, then immediately spring back to life after the first rainfall. I'd press my hand against it and watch it sink into the dense cushion. I imagined what it would be like to be an insect, nestled within the leaves, away from harm from predators, gazing up at the sky through the foliage. If you closely observe the miniature structure of moss, you'll see that it's a tiny forest for these miniscule creatures, which not only provides a home for many insects and molluscs, but on a microscopic level, tardigrades, nematodes, and protozoa inhabit the dense foliage, too.

Mosses are small, nonvascular plants that belong to the family of bryophytes and are found in various habitats around the world, such as forests, wetlands, and even the Arctic tundra. They are simple, ancient plants, and in contrast to vascular plants they don't have a network of tubes to transport water and nutrients. Instead, they have a thin structure called a phylloid, which performs photosynthesis, and their leaves contain "pores" that directly absorb moisture from rain or the humidity of the immediate surroundings. Because they lack a root system and vascular tissues, most species prefer damp environments.

Mosses are often perceived as low-light–loving plants, but this can be misleading, as there are thousands of mosses that need varying amounts of light. For example, tamarisk moss (*Thuidium tamariscinum*) struggles in the higher range of light, while mood moss (*Dicranum scoparium*) struggles in the low end. In fact, measuring the light where mosses grow in woodlands reveals they sit in areas from 180 to 1,000fc of light! A huge difference.

However, mosses will etiolate in lower light conditions and discolor in higher light, and they rarely perform well in fully sealed terrariums, quickly dying in the stagnant, constantly humid environment. Thus, maintaining an appropriate level of light, airflow, and humidity is key.

Finally, there are pH considerations. Where I live, in London, our tap water is hard and leaves a strong calcium stain on the plants and glass. This isn't ideal when watering plants but it's even less so with delicate mosses. So when watering mosses I highly recommend using distilled, deionized, reverse osmosis, rain, or bottled water (see page 35).

Mosses can range from a few millimeters to over 20in (50cm) in height—in this book we will be focusing on smaller mosses that fit into a terrarium.

Aquatic mosses

These mosses grow with most of the plant out of water, and they require a consistently moist yet airy substrate, which can be improved by adding a layer of lava rock or other suitable bonsai substrate over the soil. Aquatic mosses can also climb surfaces such as the background of a terrarium or even the glass, as long as they are moist enough. Unlike terrestrial mosses, aquatic mosses don't tolerate drying out, so they require more misting. These mosses propagate easily from the tiniest part of a leaf, and can spread across an area within months, which makes them a good, sustainable choice in your terrarium. Many aquatic mosses come in a propagation gel that should be removed before planting.

Taxiphyllum 'Spiky Moss'

PLAGIOMNIUM AFFINE

I was excited to see the creator Benji Plant had made a terrarium using *Plagiomnium affine*, or pearl moss, in his bookshelf-terrarium project on YouTube. As time has passed, the moss has colonized the entire surface of the substrate and created a beautiful, rolling-hills effect. Cutting it into smaller pieces is best, as it will spread as time goes on. It needs a minimum of 200fc of light, high humidity, and doesn't tolerate drying out. A tricky species to keep—but beautiful.

TAXIPHYLLUM BARBIERI

A commonly used species in the aquarium hobby, *Taxiphyllum barbieri*, or Java moss, has a vertical growth habit and needs good light to remain compact (200fc)—I kept a small jar on an east-facing windowsill and it grew perfectly! As it grows quickly it's important to stay on top of pruning it; a curved pair of scissors is helpful, especially in narrow containers. Other *Taxiphyllum* species that work in a terrarium include *Taxiphyllum* 'Taiwan Moss' or *Taxiphyllum* 'Spiky Moss.'

VESICULARIA MONTAGNEI

I love this species of moss! Also known as *Vesicularia dubyana*, Christmas moss is my favorite to use in a terrarium, with its horizontal growth habit and bright green color. It doesn't tolerate drying out and prefers to grow on a damp but not saturated surface—a bed of bonsai medium or a wet branch is perfect. Provide a minimum of 150fc of light and ensure sufficient airflow for it to thrive.

Terrestrial mosses

Terrestrial mosses are varieties that grow on land, as opposed to aquatic or epiphytic settings. They can be grown in any of the terrarium substrate mixes on page 31, but they do require good-quality water. All need good humidity and airflow, otherwise they will die. Mosses are very good at telling you when something is wrong: the leaves will shrivel up or discolor, or in the worst cases you will spot an odor. These are species that I've used successfully in terrariums.

DICRANUM MAJUS

Greater fork moss is a truly stunning species, with its delicate, light green leaves that are soft to the touch. It's a tricky species to keep as it grows fast, so it needs occasional pruning and good light or it'll etiolate quickly.

DICRANUM SCOPARIUM

This is another popular species that is also known as mood moss, which is apt as I find this more difficult to care for than *Leucobryum glaucum* (see right). It needs stronger light or it will etiolate quickly, sometimes brown, then even die. Native to the Northern Hemisphere, this moss covers the soil of moist forests. It's easily recognizable, with its loose, brushlike leaves that grow in an erect manner.

HYPNUM CUPRESSIFORME

Known as carpet moss, this moss is true to its name and forms a dense carpet in bright conditions. It needs good airflow and handles drying out and dormancy well. It's a popular species that covers large areas quickly in a terrarium.

LEUCOBRYUM GLAUCUM

Otherwise known as bun moss or cushion moss, this is one of the most popular mosses used in terrariums. This moss comes in dense, swollen "cushions" that can be split into smaller pieces for planting. The beige lower part can be removed, too, leaving only the green leaves, which makes it easy to plant in smaller spaces. *Leucobryum glaucum* tolerates periods of drought, simply turning lighter in color when it's dry, and will grow in a wide range of light levels. It's a temperate species but it does do well in a terrarium that is closer to a tropical environment. In nature, this species tends to grow in coniferous

Dicranum majus

Leucobryum glaucum

woodlands where the acidic tree needles fall and decompose to create an ericaceous environment, which it likes. While it's a forgiving moss, I highly advise against watering with tap water and recommend using distilled, deionized, reverse osmosis, or rainwater.

Phyrrobryum dozyanum

MNIUM HORNUM

Also known as thyme moss, this needs high humidity, good airflow, and doesn't tolerate drought. It is a good beginner species.

PHYRROBRYUM DOZYANUM

This beautiful moss species is native to China and Korea. It can be used as individual stems in smaller terrariums and looks particularly effective when they are grouped together. A taller species that requires moderate light, high humidity and ample airflow, it is simply stunning.

THUIDIUM TAMARISCINUM

Named tamarisk moss because it resembles the leaves of the tamarisk tree, this highly textured moss needs high humidity and doesn't tolerate drying out. It also needs a shady area, in light measuring 100–150fc.

Collecting moss from the wild

With the increase in popularity of terrariums, I worry that beginners will head out into the woods and take large patches of moss that have taken many years to form. Mosses play important roles in ecosystems; they help retain moisture, stabilize soils, and, importantly, provide a habitat for insects, small invertebrates, and microorganisms. These organisms rely on mosses for food, shelter, and breeding grounds, and removing moss from the wild heavily disrupts the habitat and jeopardizes the survival of these animals.

It can also land you in trouble; in many regions, removing moss from the wild without proper permits or permissions is illegal, and conservation organizations and authorities have regulations in place to protect natural habitats. So explore alternative options for obtaining moss.

Nurseries and specialized suppliers can provide sustainably sourced mosses that have been cultivated or propagated without harming natural habitats. Moss Clerks, based in Scotland, provide an ethical way to purchase mosses, as they only harvest from areas that are being felled for property development or timber production, and they also propagate their own mosses in specialized greenhouses.

Aquatic plants

When using aquatic plants in a terrarium, it is important to choose those that are slow growing, so it will take a long time for them to outgrow the container and they won't demand as much maintenance. Many aquatic species take well to growing terrestrially, which means they can be used in other types of terrariums too (see page 13), grown mounted on a piece of rock or bark, and either totally or partially submerged or positioned above water level.

In my experience, I've found that aquatic plants respond well to containers that allow for good airflow, as they dislike being in stagnant environments, and the majority are tolerant of lower light levels.

When people first create aquariums they usually begin with what is known as a dry start, which refers to the method of setting up and establishing a planted aquarium without initially filling it with water. During a dry start the aquarium is put together with a substrate, hardscape (such as rocks or driftwood), and plants, but the water level is kept very low—the substrate is just moistened but not saturated with water. This method is primarily used in planted aquariums where carpeting plants, such as foreground plants, are desired to create a lush and dense growth.

The dry start method takes advantage of the high humidity within the aquarium to keep plants hydrated while allowing them to efficiently absorb carbon dioxide from the air. This concentration of carbon dioxide is higher than what can be achieved in a fully submerged aquarium, which encourages faster and denser growth.

Creating a list of every aquatic species suitable for terrariums would be a gigantic task—far too big for this book! So I'm only going to include my favorite species here.

UTRICULARIA GRAMINIFOLIA

This carnivorous aquatic plant is native to the tropical regions of Southeast Asia and India and is often used as a carpeting, foreground plant. Its beautiful, compact, light green foliage creeps across the substrate and hardscape. It requires stronger light to grow densely and needs high humidity in a terrarium. It's important to keep the substrate far wetter than for most terrarium plants, so choosing companion plants to match this is important. Once established, it will grow over rocks or wood as long as they are sufficiently moist. It is important to provide good airflow within the terrarium, or the plant has a habit of "melting."

OPPOSITE *Utricularia graminifolia*

Anubias

This is a genus of aquatic and semiaquatic plants native to tropical central and western Africa, which grow in rivers, streams, and marshland. They are characterized by broad, thick, dark leaves that come in a huge variety of forms—from the tiny *Anubias pangolino,* with leaves no bigger than a few centimeters long, to *Anubias gigantea*, whose leaves grow to over 12in (30cm). *Anubias* are often thought of as the easiest plants to grow in aquariums, and thankfully that also translates to terrarium growing. When planting *Anubias*, it's important that the rhizome is kept out of the substrate or it will rot, causing the plant to die. These grow best in lower light conditions (100–150fc) and when attached to a branch or rock (see page 93).

Anubias barteri 'Mini Coin'

ANUBIAS BARTERI 'MINI COIN'

This rare variety of *Anubias* has tiny leaves and is extremely slow growing, making it perfect for use in small terrariums or to create a sense of scale in larger builds. It can take many months to form a new set of leaves, and each leaf grows no larger than ½in (1–2cm). The plant itself has a maximum height of 1–2in (3–5cm) and needs to be grown epiphytically rather than terrestrially. It can tolerate a range of conditions as long as the terrarium stays humid and has a good degree of airflow. Some other species of *Anubias* that work well in a terrarium include *Anubias nana, Anubias nana petite, Anubias bonsai,* and *Anubias pangolino*.

Bolbitis

Bolbitis is a genus of aquatic ferns often used in freshwater aquariums. The most popular species within this genus for aquarium usage is *Bolbitis heudelotii*, known as the African water fern, but my favorite is *Bolbitis heteroclita* 'Difformis,' due to its small size and extremely slow growth habit. This less common variety grows best attached to a branch or rock with no growing medium, because its rhizome will rot if buried in substrate. If you're patient, propagating is easy; simply snip off a piece of the rhizome and attach it to a new piece of hardscape material.

BOLBITIS HETEROCLITA 'DIFFORMIS'

A versatile miniature fern, *Bolbitis heteroclita* 'Difformis' can thrive both in a terrarium and in an aquarium. Native to the Philippines, it will grow happily in lower light levels, of around 100–150fc, but won't tolerate direct sunlight at all. It reaches a maximum height of around 6in (15cm). I've found it responds particularly well to being mounted on tree fern fiber or other porous hardscape material. It doesn't tolerate drying out and requires consistent high levels of humidity.

Bucephalandra

Bucephalandra, often called *Buce* for short, is a genus of rheophytic plants native to Borneo. A rheophytic plant is one that grows on the surface of a rock in a fast-moving stream, so it is emersed during the dry season and submersed during the rainy season, which makes this plant suitable for use in both aquariums and terrariums.

There is great variation in the foliage, from light to dark greens, to varieties with purple and even iridescent sheens under certain light. They also produce a beautiful bloom that resembles a miniature peace-lily flower! *Bucephalandra,* like *Anubias*, performs best under lower light levels (100–150fc) and should be planted onto rocks, because if they are planted in substrate the rhizome will rot, causing the plant to die. These plants have become popular in recent years, resulting in people taking them from the wild, which has decimated population sizes. Always buy from a reputable dealer who practices sustainable propagation. There are many more *Bucephalandra* species that can work in a terrarium, so try out different species and see what works for you.

BUCEPHALANDRA CATERINA

This pretty plant is easy to take care of as long as humidity is high and airflow is good. Like all *Buce*, it requires low light (100–150fc) and needs to be attached to a piece of rock or bark to grow well.

Hydrocotyle

Commonly referred to as water pennyworts, *Hydrocotyle* is a popular genus of plants used in aquascaping. They are found in wet habitats such as marshes, pond margins, and along streams. They are characterized by round, slightly kidney-shaped leaves, often borne on long stems. *Hydrocotyle* like their roots to be saturated in good-quality water (no hard tap water) and tolerate low- to mid-light ranges, so try to choose appropriate companions. Particular favorites are *Hydrocotyle tripartita, Hydrocotyle verticillata,* and *Hydrocotyle vulgaris.* For a fun project involving *Hydrocotyle* plants, check out @theplantrescuer on Instagram.

Bolbitis heteroclita 'Difformis'

Bucephalandra caterina

Aquatic plant terrarium

MATERIALS

Superglue
Large piece of lava rock (to attach the plants to)
Bowl filled with water
Bioloark container or similar with a lid
Aquarium substrate
Pressurized spray bottle filled with distilled or deionized water
Fine-grained lava rock (for top-dressing)
Long-handled tweezers

PLANTS

Anubias barteri 'Mini Coin'
Bucephalandra caterina
Vesicularia montagneioo

NOTE The container I used is from Bioloark, which has a hole for ventilation and is perfect for these plants as they require a good amount of airflow and humidity to thrive.

Planting a terrarium that totally changes in appearance within a few months can be frustrating and disheartening. That's why the aquatic plants that I've chosen here are extremely slow growing and so will take many years to outgrow the container.

I used some lava rock as the centerpiece of this terrarium, with the plants attached to it using superglue. Not only does the lava rock look good, it provides the perfect surface for the plants to grow on. The *Anubias* and the *Bucephalandra* grow from a rhizome and need to be planted above the substrate. The lava rock is porous and will allow the roots of the plants to grow on and into it, remaining moist but never saturated. It's very satisfying watching the roots grow into the lava rock!

I used aquarium substrate as the main growing medium, which is topped with lava rock. I've found *Vesicularia* moss responds well when grown on it! The moss will be the fastest-growing plant in this terrarium; when it's time to prune, keep any of the moss trimmings and add them to a propagation box (see page 212).

CONTINUED →

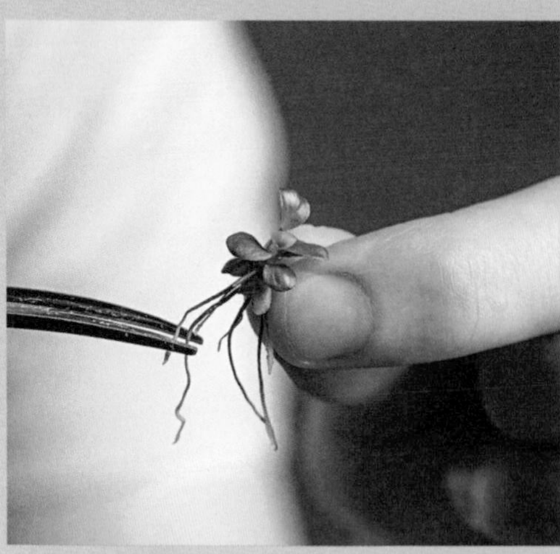

1. Remove the *Anubias* from their pots and trim the roots back until they are roughly 1–2mm in length.

2. Add small blobs of superglue to the piece of lava rock where you want the plants to go, then attach the *Anubias* and *Bucephalandra* cuttings, making sure the leaves face outward. Be careful not to glue your fingers! Leave to set. Once set, place the lava rock into a bowl of water and put to one side.

3. Place the container on a sturdy surface and remove the lid. Add a ¾in (2cm) layer of aquarium substrate into the base, arranging it on a gradient so it's higher at the back and lower at the front. Spray the substrate with water until it's damp, but make sure there is no pooling at the bottom of the container. Wipe down the sides of the container to remove any water spray or dirt.

4. Place the container back on the table and carefully firm the lava rock, with the attached *Anubias* and *Bucephalandra,* into the substrate.

5. Once it is in place, scatter the smaller lava rock pieces over the open areas of substrate. You don't have to cover every part, leaving some areas open looks nice.

6. Remove the *Vesicularia montagnei* from its pot, divide it into smaller pieces, and using tweezers place it on and around the lava rocks. Spray the terrarium with water once more so the surfaces are wet, then put the lid on top and your aquatic terrarium is finished! Keep the terrarium humid and out of direct light and it will require little maintenance—the moss grows faster than its plant companions and may need a trim occasionally.

Selaginella

Selaginella is a genus of highly unique plants that have characteristics of vascular plants, mosses, and ferns. The genus is comprised of approximately 700 species of plants that bear a resemblance to club mosses such as *Lycopodium*. Nevertheless, these two groups exhibit distinct differences. *Selaginella*, or spike mosses, possess minute leaves with scalelike flaps known as ligules on their lower surface, and they produce two types of spores. It is worth noting that common names can be misleading; spike mosses are not actual mosses as they possess a vascular system; they are capable of growing upright and significantly taller than mosses and their relatives.

Selaginella can be found in various regions across the globe and has adapted to a wide range of climates. While most species thrive in the tropics and subtropics, they can also survive in temperate regions, with even a few species found in the Arctic. The majority of species require moist soils within their natural shaded forest habitats. However, a few species, such as *Selaginella lepidophylla*, are native to the deserts of Chihuahua, Mexico, and can endure years without water. In fact, this particular species sprouts new green leaves after rainfall, and so is known as the resurrection plant due to its apparent ability to revive from a seemingly lifeless state.

Spike mosses are cultivated for their foliage, which presents a variety of shades of green, including cultivars with golden or variegated leaves. Notable species, such as *Selaginella uncinata* and *Selaginella willdenowii*, exhibit iridescent blue foliage attributed to intricate structures on the upper surface of their leaves. Species with underground runners are often employed as ground cover, but they can become invasive if grown in their optimum conditions. For instance, *Selaginella kraussiana*, a common species found in greenhouses, is prohibited from sale and commercial propagation in New Zealand due to its invasive tendencies.

In my experience, *Selaginella* is one of a few plants that do best under low light conditions (100fc or even less)—in fact, too much light and they discolor and quickly throw a tantrum. I've left a jar with *Selaginella uncinata* in a dark corner for months and when I came to look at it, it had an intense iridescent glow and looked incredibly healthy!

SELAGINELLA ERYTHROPUS

Native to Costa Rica and tropical South America, *Selaginella erythropus* is a stunning plant with a deep purple leaf and a vibrant ruby red color to the underside of its leaves. It requires a consistently damp, well-drained substrate and high humidity for it to thrive. It doesn't like intense light and should be kept away from any amount of direct sun.

SELAGINELLA UNCINATA

Perhaps the most well known of all selaginella, *Selaginella uncinata* is native to China and has a striking, iridescent blue coloring to it when it is placed in the right light. A forest-floor plant that is adapted to intense shade, it requires moist soil and high humidity to thrive. In optimum conditions it will grow quickly in your terrarium and will require frequent pruning. Thankfully, those cuttings take easily and can be used in other builds! A beautiful plant.

SELAGINELLA WILLDENOWII

Native to Southeast Asia, *Selaginella willdenowii* is another iridescent selaginella with a slightly metallic blue/green/pink hue to the leaves, depending on the angle of the light. If it is given space and not pruned it will grow large, although varieties vary in size. Like all selaginella, it needs a moist substrate and high humidity for it to thrive, and because it is adapted to grow on the forest floor in intense shade, make sure it doesn't sit in any amount of direct sun. Bright light will make the leaves dry and curl.

Selaginella uncinata

Selaginella willdenowii

Pilea

Surprisingly (to me, at least), the genus *Pilea* sits within the nettle family (*Urticaceae*). Most are native to tropical and subtropical regions around the world, with plants often characterized by their unique and attractive foliage that comes in various shapes, sizes, and textures. Pilea vary widely in appearance—some species, like *Pilea glauca* or *Pilea microphylla*, have small, round leaves, while others, such as *Pilea peperomioides,* have pancake-shaped leaves at the end of long petioles.

Pilea plants are relatively easy to care for, with many species able to thrive in indoor environments and terrariums when planted in a well-draining medium. For compact growth, make sure they have as much light as possible; when grown as houseplants they tolerate some direct sun but when planted in a terrarium this can cause a damaging greenhouse effect. They propagate easily through cuttings.

PILEA CADIEREI

Native to China and Vietnam, *Pilea cadierei* has beautiful dark, glossy leaves with intricate silver patterning that almost resembles a watermelon. Plants can become vigorous and so will often need pruning back, but it is easy to take cuttings from them.

PILEA GLAUCA

Much confusion hovers around this plant's nomenclature, so you may see it listed as *Pilea libanensis*, *Pilea glauca*, *Pilea aquamarine*, *Pilea glaucophylla*, red-stemmed *Pilea*, grey artillery plant, *Pilea* greyzy, or silver sparkle *Pilea*. Whatever it's called, it's one of the most reliable terrarium plants, and if given enough light it will form a mat of silvery blue leaves that can cascade downward when planted at a height. In lower light the distance between leaf nodes elongates and it grows upward in search of light.

Stems can be a vibrant red in color in standard light conditions. I much prefer to plant *Pilea glauca* as rootless cuttings rather than as a whole plant, as its small leaves work well in a variety of terrarium sizes and add a degree of interest to the overall aesthetic. Cuttings almost always take, and one mature plant can provide enough cuttings for many, many terrariums.

PILEA HITCHCOCKII

Native to central Ecuador, *Pilea hitchcockii* has an upright growth habit that can reach 8in (20cm), and unusual leaves that have a red, olive, and dark purple hue to them. Like most pilea, it takes to cuttings easily and performs best in a humid terrarium with bright light. Lower-light environments cause the plant to etiolate.

PILEA MICROPHYLLA

This tiny-leaved pilea has a creeping growth habit. It won't tolerate drying out, and if you want compact growth you should ensure the plant receives at least 200fc of light.

PILEA MOLLIS

One of the larger-leaved pilea, also known as *Pilea* 'Moon Valley,' this is native to Central and South America and is better used in larger terrariums. It has thick, fleshy stems that can reach 12in (30cm) in height without pruning, and crimpled, light green leaves with purple patterning. It's not one of my favorites to use, but its popularity warrants it a place on this list.

PILEA PINOCCHIO

A stunning, rare species of pilea and one of the smallest-leaved I've seen, this plant has a compact growth pattern that in time slowly creeps and acts as ground cover. It has the same requirements as most terrarium-suitable pilea: bright light (200fc) and a moisture-retentive, well-draining substrate that isn't allowed to dry out.

Pilea cadierei

Pilea glauca

Peperomia

I could have written about so many species of *Peperomia,* but I limited myself to just seven (this wasn't easy). It's a huge genus of plants, with many beautiful varieties suitable for use in a terrarium. *Peperomia* are native to many tropical and subtropical areas of the world, from Africa to Southeast Asia, to Neotropic America. Some species have large leaves and grow terrestrially, while others are tiny leaved and grow epiphytically. I'm a much bigger fan of the smaller-leaved, epiphytic varieties! Terrestrial plants tend to be shallow-rooted and detest sitting in a wet substrate, and most species enjoy high humidity.

The flowers are unremarkable and spikelike in appearance, emerging in the summer, which is why *Peperomia* are always grown for their leaves. In fact, many people cut off the flowers—myself included.

PEPEROMIA ANTONIANA

This is a species that you won't see in garden centers or plant shops. There are two kinds of *Peperomia antoniana*—one with metallic red leaves that are peppered with silver specks, the other with light green leaves and silver specks. This is a terrestrial variety that has a bushy growth habit. Under good light it will form small leaves and have a more compact growth (aim for 200fc). Make sure it has a very well-draining medium and is grown in a humid terrarium that has adequate airflow.

Peperomia antoniana

PEPEROMIA EMARGINELLA

This is one of the smallest-leaved *Peperomia* I've ever seen! Native to the Caribbean, Brazil, and Peru, this miniature epiphytic variety thrives in terrariums of all sizes. It grows best under 150–200fc of light and in a humid environment. Place cuttings directly on a bed of moss, a branch, or well-draining substrate.

PEPEROMIA HISPIDULA

A truly beautiful plant! Another small-leaved *Peperomia* that's native to Mexico and tropical America, it has oblong-shaped leaves peppered with small, white polka dots. Stems are delicate and the plant seems to tie itself into knots as it grows. This dainty plant does well in a light range of 150–250fc.

PEPEROMIA JAMESONIANA

This rare, epiphytic *Peperomia* has a creeping growth habit. Leaves are teardrop shaped with a silver marking down the middle. It does best under indirect light at around 150–200fc, becoming etiolated and messy under low light.

Peperomia hispidula

Peperomia emarginella

PEPEROMIA PROSTRATA

An epiphyte native to the rainforests of Brazil, *Peperomia prostrata* is a fairly common houseplant in Europe, but when used as a terrarium plant, it really shines. *Peperomia prostrata* is a low-growing epiphyte with small, round, succulentlike leaves with a striking pattern that resembles a turtle's shell, which is where it gets its common name, string of turtles.

When I first started making terrariums, I thought this plant was a succulent—and thus an arid-region plant that would not survive in high humidity—so I actively avoided using it for many months. It was only when I placed a few cuttings into a paludarium that sat at over 95 percent humidity that I saw it truly thrive. The leaves swelled up and the plant quickly pushed out white roots that gripped the wet bark. It was incredible to see. It thrives in bright light—a range of 150–200fc is sufficient—but won't tolerate much direct sun.

For the best results, plant *Peperomia prostrata* epiphytically in a humid terrarium. As it is a shallow-rooted epiphyte it detests overly wet substrate and will quickly rot if sat in it for too long; in a simple terrarium, plant it directly on the surface of some moss. This plant propagates easily through cuttings, but rest the rootless cuttings on a damp surface of moss, rather than substrate.

Peperomia prostrata

PEPEROMIA ROTUNDIFOLIA

Another commonly available species, *Peperomia rotundifolia* is an easy-to-keep terrestrial species with light green, circular leaves that sit on erect stems. It propagates easily through cuttings and grows quickly in a humid environment as long as it's in a suitable, well-draining substrate—it detests overly wet soil. Peperomia are generally forgiving of lighting conditions, but for optimal growth make sure it receives a minimum of 150–200fc of light.

PEPEROMIA VERTICILLATA 'RED LOG'

A common terrestrial species, but one with striking features, the leaves grow fairly large and have a dark green upper side with interesting veining and a deep red underside. Juvenile leaves are tiny and have even more veining on them. It does well in high humidity and will push out roots at leaf nodes, but it detests wet soil and needs a well-draining, water-retentive mix. It is tolerant of lower light levels, but for compact growth and attractive color, ensure it receives a minimum of 150fc of light.

Marcgravia

Marcgravia is a genus of terrestrial vines that are simply stunning in appearance, so it's no wonder they are so sought after, with some species commanding high price tags. These plants were named after naturalist George Marcgraf, who first saw them on a voyage to Brazil in the early seventeenth century. The genus includes approximately 60 recognized species, which are primarily found in the tropical regions of Central and South America. They are some of the most beautiful and desirable terrarium plants, with the vast majority being epiphytic in nature. Provide them with a background to climb and you will see these wonderful plants at their best.

Marcgravia are referred to as shingling plants, due to the way the leaves overlap and sit flush against the surface that the plant is climbing on. The leaves vary in size and shape depending on the species, but often have distinctive patterns, venation, and colors, which add to their aesthetic appeal.

This genus is commonly found in the understory layer of tropical forests, where they grow epiphytically on tree trunks or branches in sunlight that filters through the dense canopy. They are simple plants to keep and must have high levels of humidity to grow.

MARCGRAVIA SP. 'BRONZE'

A gorgeous dark and rainbow-hued species from South America, this was once considered a rare species but it has become cheaper and more available in recent years. It is very easy to grow as long as it is in a good light range with a minimum of 100fc and humidity of at least 80 percent.

MARCGRAVIA SP. 'SMALL ROUND'

My favorite, this is a delightful species with tiny, round juvenile foliage that matures to become larger and velvety. Native to Ecuador, it is an easy plant to grow in a terrarium.

MARCGRAVIA SINTENISII

Often called Sintenis' Marcgravia, *Marcgravia sintenisii* is native to the rainforests of Central and South America. It has beautiful dark green leaves with a glossy, velvety texture, and is a vigorous grower with a climbing habit. It's relatively easy to care for and thrives in a humid, warm environment that has adequate airflow. It propagates easily via stem cuttings.

Recommended *Marcgravia*

Other species of *Marcgravia* I recommend checking out include:

- *Marcgravia* sp. 'Copper'
- *Marcgravia* sp. 'Dark Brown'
- *Marcgravia* sp.'Melon'
- *Marcgravia calcicola*
- *Marcgravia umbellata*

Marcgravia 'Mini White Line'

Marcgravia umbellata 'Red'

Begonia

Begonia is a diverse and large genus of flowering plants, which comprises over 1,800 species and numerous cultivars. It belongs to the family *Begoniaceae* and is native to tropical and subtropical regions around the world, with a concentration in Central and South America, Africa, and Southeast Asia. Begonias are highly popular as ornamental plants due to their attractive foliage, colorful flowers, and adaptability to various growing conditions.

The genus contains a number of beautiful species that are suitable for use in a terrarium, many of which have similar care requirements. Many begonias like high humidity but tend to "melt" if they are in a stagnant environment, so it's important to provide sufficient airflow. Using one of the terrarium mixes on page 31 will give them the perfect substrate to root in, and it is important it is not allowed to dry out. Typically speaking, begonias prefer to be out of harsh sunlight, and they thrive in a light range of between 200 and 400fc, but you should always research whichever species you want to use, as this may differ.

Propagating begonias is easy; take a stem cutting with 3–5 leaves on it and dip the bottom end in a container of water; the cutting will release a rooting hormone into the water and in a few weeks will have produced roots. Getting the cuttings quickly into a medium such as sphagnum moss is advisable, as this improves the chances of success—transferring the cuttings directly from the water into a terrarium reduces the success rate. Once the cuttings are positioned in the sphagnum moss, place the pot into a propagating tub to allow the begonia to establish a substantial root system.

The species of *Begonia* I have recommended here are usually grown for their stunning foliage, but they all produce beautiful small flowers, too, which are a welcome surprise in the world of terrariums.

BEGONIA BOWERAE 'TIGER'

A more common species but certainly one that captures the eye, this is a rhizomatous species that has different cultivars within it, but it is a popular choice for terrariums due to its small leaf size and vigorous growing nature. It can reach heights of up to 12in (30cm), but it can be pruned and kept smaller. Commonly called the eyelash begonia because of the small hairs that sit on the outside of the leaves, it's a simple plant to keep in a terrarium as long as the humidity is kept high and the airflow is healthy.

BEGONIA CHLOROSTICTA

Now, this is a truly stunning plant. *Begonia chlorosticta* is a shrublike begonia that grows on damp cliffs and rocky slopes in its native Malaysia. There are numerous forms of this variety that come in different colors, but the leaves on this plant are olive green with a lighter green that polka-dots each leaf. It grows up to 12in (30cm) tall and blooms with interesting white flowers. It can be a difficult species to grow, but I've found that high humidity combined with good airflow gives it the best chance of success. Even with this, it can melt for seemingly no reason.

It requires a well-draining medium that is allowed to dry out a little between waterings but, just to confuse things a little more, it's also sensitive to overwatering.

BEGONIA DODSONII

This is one of my favorite begonias, not only because it looks stunning, but because it's also easy to grow! An epiphytic species from Ecuador, it has greenish-maroon leaves that are covered in fine hairs. It does well in moderate, indirect light and high humidity. It does not like water sitting on its leaves, so be careful when spraying. It takes to cuttings easily and can be placed directly into a terrarium with a high success rate. It will cover a background if given the chance, and is prolific once established.

BEGONIA DRACOPELTA

Recently described in 2019, *Begonia dracopelta* is native to Borneo, where it grows at the base of waterfalls in a very humid environment. The leaves are mid-green and decorated with dark bumps that give this species its unique appearance, and its flowers are white with a faint pale-pink tint. It needs constant high humidity and good airflow.

Begonia dodsonii

Begonia polilloensis

BEGONIA ELAEAGNIFOLIA 'SCHULZEI'

This was one of the first "rare" plants I purchased and it's been a favorite of mine ever since. An epiphytic species, this is native to Africa, growing from Cameroon to the Congo. Its small leaves, beautiful tiny pink flowers, and climbing growth habit make it perfect for use in terrariums of all sizes, and despite it being a rarer species, it's one of the easiest to keep. It does best in high humidity but is tolerant of lower levels.

BEGONIA POLILLOENSIS

A unique and rare tropical species native to the Philippines, this begonia has thin, palmate leaves that are almost fernlike. The leaves are bright green, and grow beautifully from red stems. It produces interesting pink flowers and is easy to propagate via water cuttings (see page 208).

BEGONIA SIZEMOREAE

Native to Vietnam, this can be a tricky variety to look after, but with enough humidity, airflow, and an evenly damp substrate it should grow happily in your terrarium. The leaves are different shades of green and crinkled in texture, and it has interesting long hairs on its foliage that are unique to this species. Its flowers, in two shades of pink, are produced in late summer and fall. When this plant is grown in low light it seems to develop an iridescent blue tint.

BEGONIA VANKERCKHOVENII

This is a rare but easy-to-keep species of begonia from West Africa that blooms consistently with small yellow flowers with an orange/red tint at the bottom of each. It requires high humidity and an evenly damp substrate, and doesn't tolerate drying well. I've found it needs bright light (200fc or more) for compact growth and to flower well.

Begonia sizemoreae

Begonia vankerckhovenii

Fittonia

Fittonia are among the most commonly available terrarium plants, with nearly every plant shop and garden center stocking them. They come in a huge range of colors, and during workshops people always seem to choose them first! Native to the humid rainforests of South America, *Fittonia* come in a variety of leaf shapes, colors, and patterns, but all are covered with a unique veining pattern that gives them their common name of the nerve plant.

While they're often considered problematic houseplants, due to their habit of drooping when underwatered or melting when slightly overwatered, the conditions within a terrarium make them one of the easier plants to care for. Give them an airy but water-retentive medium (such as the terrarium mixes on page 31), ensure they have high humidity, a minimum of 150fc of light, and don't sit them in a fully sealed container.

I have found that *Fittonia* respond well to cuttings, which is my preferred method of planting them in a terrarium. Instead of populating a terrarium with larger whole plants, it's often more aesthetically pleasing to integrate a few cuttings among a variety of other plants. However, the colorful terrarium project on page 176 is an exception to this rule!

Interestingly, despite their appeal, there is some resistance toward *Fittonia* within the terrarium community, particularly from rare-plant collectors. While certain brightly colored varieties might appear to some to be overly "loud," using these striking plants in creative ways can make them a real focal point of any terrarium.

While there is seemingly an infinite amount of *Fittonia* available, with different varietal names it can be really hard to differentiate between them and they will mostly be offered for sale simply as *Fittonia*.

ABOVE *Fittonia albivenis* 'Mosaic Springtime'

ABOVE *Fittonia albivenis* 'Mosaic Revolution'

ABOVE *Fittonia albivenis* 'Pink Tiger'

ABOVE *Fittonia albivenis* 'Red Vein'

ABOVE *Fittonia verschaffeltii* 'Forest Flame'

ABOVE A bold *Fittonia* mix, as in the terrarium overleaf.

Colorful Fittonia terrarium

MATERIALS

A glass container with a lid
Small-grained expanded clay pebbles, moler clay, or grit for drainage (enough for a ¾in/2cm layer)
Terrarium substrate (enough to fill a quarter of the container)
Pressurized spray bottle filled with distilled or deionized water
Long-handled scissors
Long-handled tweezers

PLANTS

Different-colored *Fittonia albivenis* plants

I was inspired to create this terrarium by Danijela Botaster, one of the admins in my Facebook group. It's a simple but effective way of using variously colored *Fittonia* to create an attractive terrarium that requires little maintenance. It's best to choose the smaller-leaved *Fittonia* and try to get as many different colors as possible! As we are using rootless cuttings, the plants will grow slowly while a root system is established, which means less maintenance in the short term, but as they grow and are pruned the *Fittonia* will fill out and become bushier. I recommend using a large container with a wide base so you can add lots of plants, but this planting concept works with containers of all sizes.

CONTINUED ⟶

1. Place the container on a sturdy surface and carefully add a ¾in (2cm) layer of drainage material into the base of the container, spreading it out evenly.

2. Pour in the terrarium substrate until roughly a quarter of the container is filled, then arrange it on a steep gradient, so it's higher at the back and lower at the front.

3. Carefully water the substrate until the top half is evenly damp; watch the substrate from the side while you're watering and once the top half has changed color, stop. Watering at this stage keeps the substrate in place and provides a stable anchor point for the cuttings to sit in.

4. Take cuttings of varying sizes from the *Fittonia* plants, making sure each cut is made just above a leaf node.

5. To position the cuttings, make a small hole in the substrate using the tweezers and lower in the cuttings until the first set of leaves is almost in contact with the substrate. Use your fingers or the tweezers to firm the soil around the stem. Place the larger cuttings toward the back, the medium-sized ones in the middle, and the smaller ones at the front. Try not to leave any spaces between the cuttings—these gaps can be plugged with more plants. Take your time, as it can take a while to get this looking right.

6. Place a lid on top and situate the terrarium somewhere that it'll get a minimum of 150fc of light.

Philodendron

Philodendron is a vast genus of plants, with over 500 species, which is native to the tropical regions of the Americas. I've found *Philodendron* species are tolerant of a wide range of conditions; they are not overly fussy about substrate, and any of the mixes on page 31 will keep them happy. I have even seen these plants grown on sphagnum moss. *Philodendron* are adaptable, making them a great choice of terrarium plants for beginners. Choosing a few favorite species for this genus was difficult due to the sheer number of options, but I've chosen a selection here. *Philodendron* are tolerant of lower light levels, but in a terrarium I've found they pretty much all respond well in the 200–300fc range.

PHILODENDRON SP. 'MINI SANTIAGO'

A rarer species, but one of the best for small terrariums, this variety is native to Ecuador and has rounded leaves no larger than 1¼–1½in (3–4cm) with a subtle, veined texture. It's very easy to care for and almost thrives on neglect! It needs high humidity but is flexible with light requirements. Don't allow the substrate to fully dry out, and give it a surface to climb for maximum effect. Cuttings can easily be taken from *Philodendron* plants.

Philodendron sp. 'Mini Santiago'

PHILODENDRON HEDERACEUM 'BRASIL'

Native to Brazil, this *Philodendron* has beautiful, heart-shaped, variegated leaves that exhibit a lime green to pale yellow color. For best variegation, it's important it receives a good amount of light, but it can tolerate slightly lower levels. It is particularly useful in vivarium setups for arboreal animals, but due to its leaf size, it is only suitable for larger terrariums.

PHILODENDRON MICANS

I bought my girlfriend a *Philodendron micans* for her birthday last year and it has grown incredibly well (better than the ones I tried!). I made this the feature plant in my glass-frog tank, due to its fast growth habit and medium-sized leaves, which are perfect for the glass frogs to sleep under. While this is a more common species, I love the way it looks, with its velvety-textured, heart-shaped leaves that have an almost iridescent sheen to them under the right light. It's a forgiving plant that is suitable for medium to larger terrariums. It can grow to a height of 8in (20cm), so it needs pruning in a terrarium.

PHILODENDRON PTEROPUS 'MINI'

This *Philodendron* is similar in size and growth to both the 'Mini Condor' and 'Mini Santiago' varieties in the trade. It is another elfin Ecuadorian species with much smoother leaves compared to the two aforementioned varieties.

PHILODENDRON VERRUCOSUM 'DWARF'

This is a dwarf cultivar of *Philodendron verrucosum,* also from Ecuador. It is visually identical to the new growth of a standard *Philodendron verrucosum*, just slower growing. It requires medium to high humidity as well as adequate airflow.

PHILODENDRON VERRUCOSUM 'MINI'

This is an Ecuadorian variety that has smaller leaves than *Philodendron verrucosum,* growing no bigger than 5in (13cm), making it suitable for use in medium to large terrariums. It has stunning emerald green foliage and adheres well to backgrounds or hardscape. It needs medium to high humidity as well as adequate airflow.

Philodendron verrucosum 'Dwarf'

Philodendron pteropus 'Mini'

Jewel orchids

Jewel orchids are a group of terrestrial orchids known primarily for their foliage rather than their flowers. The leaves are adorned with intricate, often iridescent, patterns of veins that come in a variety of colors. These vein patterns contrast strikingly with the dark background of the leaves, making them look jewellike. As a general rule, jewel orchids grow in lower light, in moist, humid conditions. I find they respond well to my terrarium substrates but people often grow them in straight sphagnum moss. When taking cuttings, make sure there is some root attached to the stem. Above soil level, the root on jewel orchids often looks like a fuzzy white ball and can be mistaken for mold. Cut below this root when taking cuttings.

DOSSINIA MARMORATA

I feel like I say this about all jewel orchids, but this one is truly showstopping. I love how the dark leaves look like they're patterned with luminous, mini lightning bolts. *Dossinia marmorata* is endemic to Borneo, where it grows in leaf litter in limestone-rich areas. Generally speaking, I've found this one of the easier species to care for. It takes to stem cuttings well.

GOODYERA MALIPOENSIS

This amazing jewel orchid has elongated green leaves with an interesting silver veining to it. Native to China, *Goodyera malipoensis* needs high humidity, a moist substrate, and a light range of 100–150fc to thrive. It propagates easily via stem cuttings.

MACODES PETOLA

Probably the most well known of the jewel orchids, *Macodes petola* is native to Southeast Asia, where it grows on the forest floor in low levels of light. The leaves are striking, decorated with dramatic veining throughout. It grows best in a humid but well-ventilated terrarium, away from bright lights and in a well-draining, moisture-retentive medium.

VRYDAGZYNEA TRISTRIATA

This species is native to Thailand, Malaysia, and Borneo. It grows in tropical lowland rainforests on limestone and in riverine secondary forests at elevations around 1,000–2,600ft (300–800m) as a miniature-sized, hot- to warm-growing epiphyte with a purplish-red, erect stem carrying a few pale, longitudinal-striped, ovate leaves. The inflorescence is roughly 1in (2–3cm) and is dense with many tiny flowers. It requires high humidity and good airflow to thrive.

CLOCKWISE FROM BELOW LEFT
Dossinia marmorata, *Goodyera malipoensis*, and *Anoectochilus roxburghii* 'Red.'

Low-light terrarium

MATERIALS

Microfiber cloth and dish soap
Terrarium substrate (see pages 24–30)
Glass candle holder or similar container, with lid
Pressurized spray bottle filled with distilled or deionized water
Neutral-colored bonsai medium, like lava rock
Long-handled scissors

PLANTS

Macodes petola or other jewel orchid
Selaginella uncinata

I've purposefully chosen this as a project for those who don't have a great deal of window space. Using a desk lamp with a warm white bulb or a north-facing window are the best lighting options, as these plants are both forest-floor species that perform well in 100fc, or even less—high light levels cause them to discolor. To recreate their rainforest-floor habitat, I'm using sphagnum moss, tree fern fibers, leaf mold, worm castings, and akadama to create a highly water-retentive yet open mix that the plants freely root into. I've added springtails here as rainforest floors are teeming with tiny creatures that contribute to a healthy ecosystem, and they will consume the leaf mold and help prevent any mold issues. These kinds of terrariums really excite me! The *Selaginella* has the most incredible iridescent glow and basically thrives on neglect, so the only maintenance you need to do is pruning it, as it grows fast! The *Macodes petola,* or jewel orchid, is one of the most stunning groups of foliage plants in the world and grows nicely next to the *Selaginella*.

CONTINUED ⟶

1. Clean the container using a microfiber cloth, warm water, and dish soap.
2. Add a few inches of the terrarium mix into the base of the container, or use sphagnum moss. Carefully water the substrate until it is lightly moist. Take extra care not to overwater, though, as there is no drainage layer.
3. Remove the jewel orchid from its pot, or if you are taking a cutting from a larger specimen, choose a stem with at least two leaves on it and ideally some root (roots on jewel orchids often form above soil level).
4. Make a small hole in the substrate, just big enough for the jewel orchid to fit into, and place it into the hole up to the first set of leaves. Firm the orchid in place.
5. Top-dress the open areas of the soil surface with a neutral-colored bonsai medium like lava rock (you can skip this step if you don't have any bonsai mediums).
6. Take some *Selaginella* cuttings that have a few roots forming under the leaves, then trim these roots until they're 1mm in size—this helps them find their way into the new soil. Place a few of these small cuttings around the jewel orchid and lightly spray with water.
7. Place the lid of the terrarium on, then position it in a spot away from bright light.

Ongoing care

Keep an eye on the substrate to ensure it stays evenly moist but not overly wet. Remember the analogy about the damp sponge (page 37)? That's the level of moisture to aim for. *Selaginella* can grow fast, so keep on top of it by regularly pruning so it fits in the container.

Cuttings can be taken from the jewel orchid and used either in the same terrarium or a new one. If the plants start to discolor it's a sign they're getting too much light, in which case move the terrarium to a shadier spot.

Orchids

An orchid is a flowering plant that belongs to the family *Orchidaceae*, which is one of the largest families of flowering plants. Orchids are often known for their diverse and beautiful flowers, and are highly prized for their ornamental value, being popular among plant enthusiasts, florists, and collectors alike.

Orchids are found in various habitats around the world, including tropical rainforests, temperate forests, grasslands, and even deserts. They have adapted to these diverse environments and can be found on nearly every continent. Each type of orchid has specific requirements in terms of light, humidity, temperature, and growing mediums, and not all are suited for use in a terrarium, such as, for example, the popular *Phalaenopsis* orchid.

Some orchids are epiphytic, which means they grow on other plants or objects without taking nutrients from them. Epiphytic orchids often have aerial roots that absorb moisture from the humid air. Others, like jewel orchids, are terrestrial and grow on the ground in rainforests in the understory and forest-floor layer, where they receive low levels of light and high levels of humidity. Some orchids, such as *Dendrobium* and *Paphiopedilum,* are lithophytic, meaning they grow on rocks or rocky surfaces.

SPECKLINIA DRESSLERI

A beautiful, tiny-leaved, epiphytic orchid that's native to Panama. It has tiny circular leaves that trail down, and it grows particularly well on sphagnum moss or wet wood, such as tree fern or cork bark. To me, it almost has a kind of *Peperomia* look about it. It produces large reddish-orange flowers in comparison to its leaf size, and is a rare, easy-to-grow orchid for terrariums. I've found this does well in the 150–250fc light range and is happy at room temperature.

BARBOSELLA DUSENII

This tiny-leaved epiphytic orchid is native to the cloud forests of Brazil. In a terrarium it does best when mounted on a piece of wood, and requires frequent misting as it flourishes in a very humid environment. As with most epiphytic orchids in a terrarium, it's vital there is a good amount of airflow. Even when not displaying its little yellowish-green flowers, this is a stunning plant!

Recommended orchids

I'm certainly no expert on orchids but I recommend researching miniature species within these genera. You'll find some real gems:

- *Aerangis*
- *Angraecum*
- *Bulbophyllum*
- *Haraella*
- *Lepanthes*
- *Masdevallia*

Biophytum

Biophytum belongs to the family *Oxalidaceae,* which is more commonly known because of the genus *Oxalis* that includes the wood sorrels, plants often seen in British woodlands. I remember my uncle showing me these tiny plants in Singapore when I was a child and I was fascinated by their resemblance to miniature trees! They often grow to no more than around an inch tall, and are found in wet areas in tropical countries, making them perfect additions to a terrarium. Their forgiving nature and prolific seeding habits make them a fascinating, easy choice for beginners.

Biophytum species resemble miniature palm trees, which gives them their common name of little tree plant. *Biophytum* need a substrate that will not dry out, and as they come from tropical regions they also require high temperatures. Much like the well-known *Mimosa pudica*, or sensitive plant, when the leaves of *Biophytum* plants are touched the leaflets fold together, and at nighttime they fold inward on themselves, as if they're retiring for the night. The flowers are small but can be quite attractive, often yellow or white.

When the plants produce seedpods, they burst open in a starlike fashion, propelling the seeds outward. In the wild they will colonize areas quickly due to their rapid seeding habits. You'll see this habit replicated in a terrarium, with seedlings sprouting up or even seeds germinating on the sides of the container. *Biophytum* plants often command a high price point, so it's advisable to sow and germinate seeds yourself (see page 208) rather than allowing them to self-seed within the same terrarium.

BIOPHYTUM SP. 'ECUADOR'

A stunning, darker-leaved species from Ecuador, this has slightly larger and thicker leaves than *Biophytum sensitivum*, as well as bright pink flowers that stand out beautifully against the dark leaves. In the wild it is usually found growing in marshy areas, or next to streams or waterfalls where there is constant moisture. It's a wonderful addition to any terrarium.

BIOPHYTUM SENSITIVUM

Easily the most famous of the *Biophytum* genus, *Biophytum sensitivum* is widely available and simple to grow. It doesn't like large amounts of light and sits happiest between the 100–250fc range. Ensure the substrate doesn't dry out. Plants reach 4–6in (10–15cm) in height and spread, and seedlings look especially nice when grouped together.

OPPOSITE *Biophytum sensitivum*

Bromeliads

Bromeliads are distinct in appearance, immediately recognizable by their unique rosette of leaves that stems from a central point. This rosette forms a "cup" in the center that is adept at collecting water, making them a popular choice for dart-frog vivariums. Interestingly, many bromeliads are epiphytes, which means they grow on other plants, such as trees, without leaching nutrients, instead deriving nutrients from rainwater, airborne dust, and the decaying matter that gathers in their rosettes. In a terrarium, epiphytic bromeliads are mounted directly onto a branch or background such as cork-bark or tree-fern slabs, using fishing line or superglue, sometimes with sphagnum moss attached to the roots.

It's more common to see bromeliads in larger terrariums and vivariums than smaller bottle-garden–style plantings. When choosing bromeliads for terrariums, it's essential to select species that match the care requirements of the other plants growing in the same space. Some species require high amounts of light, so this must be taken into consideration when choosing plants to partner with. While it's unwise to pair plants with widely different light requirements, sometimes the size of the terrarium means that the top of it gets strong light while the bottom gets little. Bear this in mind when choosing your plants. Regardless of what kind of bromeliad or terrarium you choose, *all* bromeliads require good drainage.

For watering, the ambient humidity in a terrarium is not going to be enough to keep them watered, so a misting system, or hand misting, is required. Bromeliads are sensitive to tap water components such as chlorine and fluoride, so it's important to use good-quality water (see page 35).

AECHMEA ORGANENSIS (SMALL FORM)

This species is endemic to southeastern Brazil. These urn-shaped plants collect water in their leaf axils, and they are a nice novelty, with upright rosettes growing to 6in (15cm). The leaves are green with darker bases and a short, sharply pointed (apiculate) tip. It's a small species, perfect for a terrarium.

OPPOSITE *Aechmea* sp.

CRYPTANTHUS BIVITTATUS

A commonly used terrestrial bromeliad from the humid rainforests of Brazil, *Cryptanthus bivittatus* is known for its star-shaped leaves in a range of distinct bright colors. Due to its leaf shape, it has the common name of the earth star plant. It is easy to care for, requiring a substrate that is kept on the drier side, and while it prefers a high humidity range, it can tolerate lower light levels.

Other *Cryptanthus* species suitable for a terrarium include *Cryptanthus fosterianus* and *Cryptanthus beuckeri.*

GUZMANIA LINGULATA MINOR

Native to Central and Southern America, this bromeliad is commonly seen in garden centers and plant stores. *Lingulata* means "tongue-shaped," and this species grows to just under 12in (30cm) tall, making it a good choice for medium-sized terrariums. Its leaves grow to about 4in (10cm) long and it produces a cup-shaped flower on a short stem. *Guzmania lingulata* is most famous for its pink or red flower bracts that adorn the stem. While generally grown as an epiphyte, a well-drained substrate can support them. Ensure it has adequate airflow and a higher light level of at least 400fc.

NEOREGELIA 'AMAZON'

This beautiful, small-growing *Neoregelia* has attractive maroon leaves patterned with dark flecks. It's simple to grow but should be mounted onto cork bark with some moss to root into. The brighter the light you give them, the more intense the coloring on the leaves becomes.

NEOREGELIA 'MO PEPPA PLEASE'

Picking just two *Neoregelia* species for this section was difficult, but with a name like this, how could I leave it out? Not only does it have a fun name, it's an eye-catching plant. A miniature bromeliad that typically reaches 6in (15cm) high or less, it grows a compact rosette of green leaves that are peppered (sorry!), speckled, with dark spots. This is a plant that does well in strong light.

Neoregelia 'Amazon'

Neoregelia 'Mo Peppa Please'

Tillandsia stricta

TILLANDSIA STRICTA

This is a common species in eastern South America, from Venezuela to Northern Argentina, where it grows as an epiphyte in dry to mesic (evenly moist) conditions. It has green, often gray/purple mottled leaves and a lovely flower spike with bright pink bracts and light purple flowers. It is popular among beginners for its appearance and ease of cultivation, and it produces many pups (offsets) and quickly grows into an attractive clump. This species likes to be in a light range of 200–400fc, and requires good air movement. It's essential to mount them on a branch or background. Keep a close eye on the plants and water as needed; typically they should be removed from the terrarium, submerged in water, then placed back on their mount, which allows them to dry between waterings. They will not grow well in sealed terrariums and are happiest when mounted in large terrariums that have good light and airflow.

Tillandsia bulbosa and *Tillandsia filifolia* are also suitable *Tillandsia* species for terrariums.

Other plants

Here is a selection of plants that belong to miscellaneous genera that are great for terrariums—perfect as both feature plants and fillers, for their foliage, flower, and form.

CHAMAEDOREA ELEGANS 'MINI'

This commonly available plant works well in the background of a terrarium. Pots usually come planted with 10–15 individual plants. It's best to split these up by separating them but still leaving some root intact. When planting, trim roots back a little to help them find their way into the new substrate. It's tolerant of lower light levels, around 100fc, but for healthy growth make sure it gets around the 200–300fc range.

CHIRITA TAMIANA

I love this plant! A friend gifted me one in early 2022, and for the best part of a year it flowered constantly. Sometimes known as *Deinostigma tamiana,* it is a highly adaptable terrarium plant that flowers wonderfully in my frog tank, but also in smaller bottle-garden–style terrariums. Ensure 200fc of light for those wonderful blooms! It propagates easily by leaf cuttings; place the stem in water and in a few weeks roots will form. Plant into a terrarium as soon as roots form.

HEDERA HELIX 'MINIMA'

Unlike its larger, less-beautiful brother, *Hedera helix*, this miniature variety has every desirable trait of a terrarium plant. It has beautiful, small leaves, it's slow growing, and is simple to care for. I bought a plant for my garden in 2015 and while it grew well, I couldn't get it to propagate. It was only when I placed a cutting into a small terrarium that I had success. Within a few days, the stem had pushed out an almighty thick white root that nestled itself against the curve of the container. I've tried this many times since and on average, 8–10 cuttings take. I highly recommend you try out this plant. Give it a minimum of 100fc and ensure the substrate doesn't dry out.

HYPOESTES

Similar in appearance to a *Fittonia*, I don't tend to use these in the terrariums that I make, but as they're a popular choice at workshops, I felt it was important to include them in this book. Plant them whole or separate them into smaller pieces with root still attached. Cuttings are unreliable.

PELLIONIA REPENS

While it's commonly called a trailing watermelon begonia, it's definitely not a begonia (they're not even in the same family!). This plant has a trailing habit with a watermelonlike appearance on its oval-shaped leaves. It propagates easily via cuttings and requires a substrate that is airy and doesn't dry out.

SAXIFRAGA STOLONIFERA 'MINI'

A recent addition, and one that has shot right up my favorite terrarium plants list. This beautiful saxifrage is much smaller than the standard *Saxifraga stolonifera*, which is erroneously called a strawberry begonia (it's definitely not a begonia). I searched for this plant for a long time and I was delighted when a company in China sent me six plugs. Seriously, if you're able to get hold of some, I advise it! This saxifrage can be divided and propagated into smaller plants and grows no larger than about 1in (a few centimeters) high and will spread, if given time. It likes a minimum of 150fc, high humidity, and decent airflow.

Chirita tamiana

Hedera helix 'Minima'

Saxifraga stolonifera 'Mini'

SINNINGIA MUSCICOLA

Fulfilling that desirable, small-leaved, slow-growing trait that we love, *Sinningia muscicola*, previously known as *Sinningia* sp. 'Rio das Pedras,' is probably the quintessential terrarium plant. They tend to grow slowly, occasionally producing trumpet-shaped, pale violet flowers. Seedpods emerge after a while, with the seeds being incredibly tiny. They're easy to germinate, though. Sit seeds on a bed of sphagnum moss and transfer to a terrarium soil mix once germinated. Ensure a minimum of 150fc and use a medium that is especially water-retentive and airy. I add a little more akadama to my regular terrarium substrate for this plant.

SOLANUM SP. 'ECUADOR'

A tiny, easy-to-grow species that I was gifted a few years back. It will produce long stems but the leaves are no bigger than about ¼in (½–1cm) in length. It does spread, but it is very easy to keep in check and fits into the tiniest of terrariums. It has very simple care requirements, and needs a minimum of 100fc of light and high humidity.

STREPTOCARPUS IONANTHUS

I've not had a great deal of luck growing African violets in a terrarium, but I know other people have. Generally speaking, a humidity range of between 70 and 80 percent is best, but they need ample amounts of airflow or they can rot. While African violets are grown for their flowers, I think the foliage is equally beautiful. Ensure they get at least 200fc of light, but ideally for them to flower 300–400fc is best.

TRADESCANTIA SP.

Made famous by the David Latimer terrarium (see page 15), *Tradescantia* is a genus of perennial wildflowers native to the Americas. A common houseplant that grows well—almost too well—in a terrarium, it needs regular pruning to keep it in check. For that reason I've only used this in a few of my terrariums.

Sinningia muscicola

OPPOSITE *Solanum* sp. 'Ecuador'

Carnivorous plants

Carnivorous plants are a unique group that have the ability to attract, capture, kill, and digest insects—and sometimes even small animals—to derive essential nutrients from them. This enables them to survive in nutrient-poor environments such as acidic bogs or sandy soils, where most other plants would be unable to obtain the essential nutrients they require for survival. They are some of the most recognizable plants in the world.

Plants like *Nepenthes* (pitcher plants) have hollow, tubular leaves filled with digestive enzymes. Insects are lured to the opening by its color or nectar, fall in, then are unable to escape because of the slippery inner surface or downward-pointing hairs. The plant then digests them using bacteria and enzymes. *Drosera* (sundews), on the other hand, have sticky trichomes (tiny epidermal hairs) on their tentaclelike structures that trap insects when they touch or land on them, then over time the tentacles close around the prey to digest it.

Probably the most famous carnivorous plant is *Dionaea muscipula* (Venus flytrap), which has hinged leaves that resemble a mouth, and when the sensitive hairs on the inner surface are touched by unsuspecting insects, the leaves quickly snap shut, trapping the insect inside.

There are also many carnivorous aquatic plants, such as *Utricularia* (bladderwort), which have small bladders that create a vacuum. When tiny aquatic creatures make contact with the trigger hairs that are near the bladder's opening, the trapdoor swings open, sucking the creature inside. Shocking! Once the prey is trapped, the plant secretes enzymes to break down the insect's body. This process releases nutrients like nitrogen and phosphorus, which the plant absorbs.

While it can be tempting to use carnivorous plants in a terrarium, there are important factors to consider; generally speaking, these plants need higher levels of sunlight compared to the vast majority

OPPOSITE *Nepenthes glabrata* x *hamata*

of terrarium plants, because when in nature they tend to grow in full sun. In terms of maintenance, while they are carnivorous and feed on insects, they do not need to eat frequently at all, only requiring a few flies each year.

Water is an important factor when keeping these fascinating plants. The tap water we consume is too mineral-rich for them and mineral-free water, like rainwater, distilled, deionized, or reverse osmosis water, is important for their growth.

Another vital factor is the substrate they're grown in. Many carnivorous plants grow in acidic, sphagnum-peat bogs, and peat has long been considered the basic compost ingredient for most species, but as it's an unsustainable product that is being phased out, it's important we use alternatives. Sustainably sourced sphagnum moss is an alternative and can be used with all carnivorous species. It's often paired with perlite for aeration. Bonsai mediums are nutrient-free and I've seen people using those to grow their carnivorous plants in. Other suitable materials you can use include: perlite, pumice, peat moss, orchid bark, sand, akadama, Kanuma, coco chips, coco coir/peat and tree fern fiber.

It's wise to choose species from warmer climates for your terrarium, as temperate carnivorous plants will need a dormancy period over the winter months. (It's for this reason that I left *Dionaea muscipula* off my list of favorite plants for terrariums.) It could be possible to create a terrarium using exclusively temperate plants, but it would need to be left outside during the winter periods for them to survive.

DROSERA PARADOXA

A species that likes warm temperatures and is endemic to Western Australia, *Drosera paradoxa* requires heat and humidity all year round. A taller species that reaches up to 12in (30cm) in height, it adapts well to terrarium life and is a relatively simple plant to keep.

DROSERA SPATULATA

One look at this plant and you'll see why this is known as the spoon-leaved sundew. *Drosera spatulata* is considered a weed among carnivorous-plant growers, due to its prolific seeding habits. One of the most widespread sundews, this is a great choice for a terrarium because of its beautiful coloring and small nature, with it reaching no more than about 1in (a few centimeters) in diameter. It spreads happily but is easy to keep in check. Ensure it gets strong light and the substrate is kept wet at all times.

NEPENTHES X HOOKERIANA

This *Nepenthes* is a hybrid of *Nepenthes rafflesiana* and *Nepenthes ampullaria* and it's nice and easy to grow in a terrarium, reaching 4in (10cm) across. It will do well planted terrestrially, but ensure the substrate is nutrient-free—it can also be mounted higher up in a pot or into hardscape. Its pitchers are green with red and purple spots, and these plants have a wide green peristome (the lip on the opening of the pitcher) and a very small, spurlike lid. This is a warm-growing species that needs a minimum temperature of 59°F (15°C).

NEPENTHES GLABRATA X HAMATA

A truly delightful plant, and one of the most stunning *Nepenthes* I've ever seen. This hybrid has many attractive qualities inherited from both its parents; the winged frontal fans of *hamata*, the beautiful painterly flecking of *glabrata*, and the globular, egglike shape of *Nepenthes tenuis*. For it to thrive, keep it warm, and the substrate moist but not saturated, and ensure it gets a minimum of 350fc of light.

NEPENTHES 'BLOODY MARY'

A nice, compact species with deep red color and dark green leaves. Unlike other *Nepenthes*, this one is less demanding and will stay in good condition, with its cups lasting for many months.

PINGUICULA ESSERIANA

This very sweet, succulentlike butterwort is native to Mexico. It grows no more than about 1in (a few centimeters) in diameter and produces beautiful pink flowers. Plant in a nutrient-devoid, open, but water-retentive medium and ensure it gets strong amounts of light. This is a good choice for an east- or west-facing window in warmer months, or a south-facing one in fall and winter.

PINGUICULA SP. 'HUATLA'

Another *Pinguicula* that's native to Mexico, this one is also small in size and has a pink hue to newer growth. It produces beautiful flowers at the end of long, woody stems, which is a real feature of *Pinguicula* in general.

Pinguicula esseriana

6

Cultivation and care

Caring for plants and ourselves

"My wish is to stay always like this, living quietly in a corner of nature." – Claude Monet

In my life, working with plants, especially within terrariums, has offered me an oasis of calm in times when I needed it most—not just during turbulent periods, but on a day-to-day basis. Working with plants has transformed my life in ways I could have never imagined. This is more than just a business, hobby, or pastime for me; there has been a profound interconnectedness between caring for plants and my own self-care.

I'm often in a state of wonder when I think about my horticultural journey and how impactful it has been on me. Taking on my vegetable patch in 2014 was a pivotal moment in my life, but I didn't know it at the time. I'd long heard about the benefits of gardening, but I'd be lying if I said I took on that plot for the mental and physical benefits it gave me, even though that turned out to be the most important aspect. Had I not got the vegetable plot I wouldn't have discovered horticulture, and had I not discovered horticulture, I wouldn't have found myself within this enchanting world of terrarium building. Life is surreal, and sometimes one seemingly small decision has a tidal wave of impact that ripples for years afterward. I can't even begin to express how happy I am that I took on, and kept, that vegetable patch.

I used to be an impatient person, and perhaps in some ways I still am, but caring for plants and these tiny worlds has required me to be patient and attentive. It's a ritual that pulls me away from the everyday chaos of life and into the present—and often into a meditative state where hours dissolve without glances at the clock. I think we all have a safe space, somewhere we feel totally comfortable and at home, and for me, it's in my workspace, listening to my favorite music while surrounded by these little ecosystems. In a world that seeks instant gratification, plants defy this norm; they grow at their own pace, regardless of our desires. When I'm around my plants, I'm subconsciously practicing being in the moment, just as in meditation or mindful breathing, and over time this not only benefits the plants but also nurtures my own mental peace.

Amid lofty goals and intangible achievements, the plants I tend—from zucchini in the garden to the tropical ferns in my terrariums, to the recent acquisition of some *Philodendron* that occupy a corner in my living room—provide me with concrete and consistent evidence of my care. That new leaf, ripe tomato, vibrant cushion of moss, or simply the survival of a once-wilting plant are the direct outcomes of my own efforts. These visible results are immensely gratifying, and the sense of accomplishment that beautifully accompanies these is something that everyone can enjoy. It's a reminder to us all that our actions have consequences and that positive effort will eventually lead to beautiful results.

As someone who loves to exercise, I've found this is a sure way to turn a bad mood into a good one, or an unproductive day into a productive one. It's something I feel strongly about, and I do my best to exercise at least four or five times a week. Let's be real, building terrariums has little in the way of physical exertion, but my time on the vegetable garden and working as a gardener or landscaper really showed me that gardening is a physically demanding activity that requires a decent level of fitness. Many people speak of the benefits that gardening has for them, both for their physical and mental health; exercise releases endorphins that relieve stress and improve your mood. You have to be consistent with exercise to see the benefits, and it's the same if you want your plants to flourish. The combination of both exercise and horticulture is a true superpower in tackling depression, anxiety, and a whole host of negative emotions, and I am living evidence of this. My time spent around plants has been like an organic antidepressant.

When I think back to the things that really changed my life, the vegetable patch fostered a strong sense of responsibility within me. It was a commitment; I made a promise to return, to check on it, and to provide. This daily routine gave me true structure at a time when I needed it most. It created a purpose, and even though it was seemingly small, that made a significant difference to my mental well-being and was the start of me turning my life around.

The past few years have been a true joy. I've had a lot of success online and had the privilege of meeting a lot of successful people, from displaying at the RHS Chelsea Flower Show, to working with Disney and Marvel, and now, writing this book. The plants in my life have grown alongside me and as I've nurtured them over the years and provided them with what they need, I realize that they, too, have provided for me.

The simple act of caring for plants extends beyond the vegetable plot or terrarium. It's a gentle walk into a mindful, patient, and joyful activity, and a journey into understanding myself better. As I look up into my workspace at the many terrariums, animals, and plants inside it, I know that while I provide them with sustenance, they reciprocate it, silently, in far greater quantities, offering lessons in tranquillity. It's a reminder that in caring for the tiny worlds, I am, in essence, caring for myself.

Propagating techniques

Despite its green and wholesome image, the houseplant industry is not very environmentally friendly. Most plants are grown in giant nurseries in Europe, often in peat-based media that are then shipped overseas in black plastic trays. The air miles, electrical costs, plastic waste, and use of peat for a single trolley are significant, and this multiplies when you consider that most plant shops buy from these nurseries. Propagating your own plants not only saves you money, it also helps reduce waste and environmental impact from buying from these suppliers.

Plant propagation is an essential skill for anyone working in horticulture, as not only does it allow you to take cuttings and grow your own plants for free, it also enables the cultivation of a wide variety of plants, including rare or unusual species that may not be readily available. I think that is the most fun part of all, as it allows you to expand your plant collection without significant expense. Successfully propagating your own plants and seeing them thrive in a terrarium enhances that sense of accomplishment and a deeper connection to the hobby.

I think of terrariums as mini propagation chambers; these glass vessels of varying sizes can be used to grow cuttings until they are large enough to be potted on or moved to another container. This perspective on terrariums shifts the focus from creating something permanent to a more creative, temporary approach. Small cuttings can be placed into homes that are aesthetically pleasing but not necessarily designed to last for many years, instead staying there until they outgrow the space and can be moved on.

RIGHT Dividing a *Nephrolepis cordifolia* 'Duffii' fern.

A common question is what to do with the plants when they grow too large. While they will need trimming, this is not a wasteful endeavor, because pruning provides us with the opportunity to get the cuttings and free plants that can be used again or even sold. Single cuttings of rarer plant species can fetch high prices, so growing them to a size where they can be sold can help offset the costs of this potentially expensive hobby.

Researching each species you buy helps you to better understand the growth habits and needs of these plants, which leads to healthier, more successful terrariums. I have learned a great deal through my own research and by talking with other hobbyists. Just the other day, I was in discussion with a member of my Facebook group who posted a picture of their propagation box (see page 212 for how to make your own), which looked beautiful. I said that I often prefer the way these boxes look to perfectly manicured terrariums and how the selection of plants in their prop box were unusual. We discussed this further and it turns out that this was their main way of keeping plants and making terrariums came second—there is something deeply satisfying about a healthy propagation box!

I also saw how a member of the *Ficus* study group propagated his *Ficus thunbergii* by growing it vertically in his conservatory, out of a terrarium. He lives in Florida, where it's quite humid, which is why he was able to grow it to such a large scale. I suppose his conservatory is basically a large terrarium!

There are numerous ways to create new plants from other plants; some techniques can be used across most plants, some are more effective for certain species.

Techniques

I'm not horticulturally trained, and everything I've learned is through research, hands-on practice, and gleaning advice from others. The terrarium hobby didn't have a whole lot of information available when I first started out, especially on propagation and the use of cuttings. I'm not saying people weren't using cuttings, but I don't recall seeing much information on this topic in the early days. Through my own experiments, I discovered that the *vast* majority of plants used in a terrarium can be propagated easily, mostly through stem cuttings but also through division, growing from seed, leaf cuttings, and even water propagation for more finicky plants like begonias.

DIVISION

Many terrestrial ferns or parlor palms don't take as stem cuttings and need to be divided. To divide ferns, remove them from the pot, find the center point of the plant and split or tear it into two equal pieces. With larger plants you can divide the root ball into multiple pieces, increasing the amount of new plants.

When it comes to planting your new plants, loosen the root ball of each and remove some of the longer roots—leave some root intact or the plant won't survive. If a lot of the root mass has been disturbed, prune out some of the foliage by tracking leaves to the base of the plant and making your cuts as low as possible, because a small root mass will struggle to support a large amount of foliage.

STEM CUTTINGS

This is a simple and easy way to propagate the vast majority of terrarium plants. You must start with a healthy plant because an unhealthy cutting is just going to struggle and is unlikely to survive.

If you're wondering what a suitable healthy plant looks like, the leaves will be strong in their color, the growth compact from adequate light, and the roots will be a healthy white color on inspection when you remove the plant from its pot. There will also be a pleasant aroma from the substrate—unpleasant smells usually come from anaerobic conditions.

To take a cutting, use a clean, sharp pair of scissors, select a stem with a few leaves on it (the cuttings can vary in size but larger cuttings with more leaves will require more stem to push into the substrate), and cut just above a leaf node (see the picture, opposite). Push the cutting down into the substrate until the first set of leaves is just above the top of the substrate. Stem cuttings thrive in a warm, humid environment, so a terrarium or propagation box (see page 212) is the best place for them! In time, the plant will produce roots and start to grow inside the terrarium.

WATER PROPAGATION

This isn't a method I use often, but it works with plants that struggle with stem propagation, such as certain begonia species. The technique is very similar to stem propagation but the rooting medium changes to water.

To water propagate, take a clear container like a jam jar or a vase and fill it with water. Take a cutting from a healthy plant that has at least 3–4 sets of leaves. Place the cutting into the water up to the first set of leaves, ensuring they are not submerged. In a few weeks the cutting will have produced roots. Once it has reached around 2in (5cm) long, transfer it to a pot of substrate or a terrarium to allow the roots to quickly adapt to soil conditions and maximize the cutting's chance of success. If you've planted into a pot, place the pot into a propagation box to increase humidity.

SEEDS

Growing from seed is something that many gardeners have experience with, but as terrarium hobbyists, it's not common to see. Propagating this way is often a more lengthy process, but it's usually the cheapest, as it's always more expensive to buy ready-grown plants than a packet of seeds. As a general rule, seeds will propagate better in a lower-nutrient substrate. Gardeners often use John Innes No.1 to start seeds off, then pot them on into a more nutrient-rich medium as the plant grows. If you make your own substrate, I recommend starting seeds off in a regular mix, but omit the worm castings. Many growers start seeds off in pure sphagnum moss, and this is perfectly fine, too. Just make sure the moss doesn't become saturated, as this will cause the seeds or seedlings to rot.

OPPOSITE Taking a stem cutting from a *Fittonia.*

PROPAGATING TERRESTRIAL MOSS

To propagate terrestrial moss, trim the green leaves, but keep a lot of the beige cushion underneath, and place them in a clear container with a lid under a good, indirect light source of around 200fc. In a few months new leaves will appear!

Interestingly, one of my Facebook group members shared that when trimming the green leaves, rather than discarding the removed beige cushion underneath, they grew on these leftover pieces in the same way.

I have only ever used this method for *Leucobryum glaucum,* but I encourage you to experiment with other mosses!

PROPAGATING AQUATIC MOSS

Aquatic species of moss can grow from the tiniest piece. When propagating aquatic mosses, I like to use akadama as the potting medium, as it stays damp while remaining airy, which is ideal for growing mosses on. Aquatic mosses need humidity and airflow, so it's important both are provided. Often, terrariums will need frequent misting, as the ambient humidity isn't enough for the moss to thrive. The mosses should never appear dry, but they should also never sit in stagnant water.

To propagate aquatic moss, take a clump and either divide it into small sections, or cut it up finely using scissors or a knife. Spread this across a surface of the akadama inside a propagation box or a terrarium, and mist it with water until it's just damp. Keep an eye on it so it doesn't dry out, and mist accordingly as needed.

LEAF CUTTINGS

Amazingly, many plants can be grown from a single leaf! *Peperomia*, begonias, and African violets can all be propagated in this way. It's not something I tend to do often, but if a *Peperomia prostrata* leaf falls in a terrarium I will just leave it and see what happens. It's quite satisfying watching the leaf melt away to leave a tiny little plant in its place.

To take a leaf cutting, remove a single mature leaf and place it directly into a terrarium, on top of the substrate or moss. Ensure the moss or substrate stays damp but not wet, and in time a new plant will appear. This isn't the most reliable way of taking cuttings from *Peperomia prostrata*, but this makes good use of the leaves that occasionally fall off.

Non-terrarium plants such as *Crassula ovata* (jade plants), *Sansevieria*, *Streptocarpus*, and *Sedum* are also easily propagated this way.

OPPOSITE When trimming *Leucobryum glaucum* moss, don't discard the beige lower part as this can be grown on to produce a new flush of moss.

How to make a propagation box

Once this hobby grips you, I've no doubt that you'll have every window space in your home filled with these miniature ecosystems. However, eventually the plants do outgrow their containers and you will need to perform a little maintenance. As the vast majority of terrarium plants take to cuttings well, and with the cost of plants being high, it makes total sense to grow them on for use later.

Propagation boxes are useful as they provide temporary homes for these cuttings. They can be filled with substrate, just like a regular terrarium, for planting, or they can be filled with cuttings in individual pots. The concept is the same. Clear plastic storage boxes are good for this purpose; they are inexpensive and work in exactly the same way as a terrarium. I recommend that you use grow lights rather than natural light in this instance, as the boxes are large and take up a lot of space. Using the grow lights enables you to place the propagation box wherever you like.

MATERIALS

Clear plastic storage tub with a clear lid
Drill with a 2–3mm drill bit
Terrarium substrate
Spray bottle with deionized or distilled water
Sphagnum moss

PLANTS

Terrarium cuttings

1. Remove the lid from the plastic tub. Drill a few holes in two sides of the tub using a 2–3mm drill bit.

2. Fill the box with a layer of substrate until it is one-fifth full. Spray the substrate with water until it's damp.

3. Add a roughly ¾in (2cm) layer of sphagnum moss across the surface to cover all the substrate, then lightly water the moss until damp but not soaked.

4. Take your cuttings and place them inside the propagation box, firming them into the moss.

Terrarium care

Much like looking after a garden or houseplant, it's fairly obvious when something is wrong in your terrarium. It's an unrealistic expectation that a terrarium will look totally healthy all of the time; some plants are fussier than others, and no matter what you do, they can throw a tantrum for seemingly no reason at all! Plants grow at different speeds and will compete with each other for light and nutrients. It's natural for faster-growing species to dominate, and if left unchecked they will eventually block out light for the other plants in the terrarium. It's also totally normal for plants to shed older leaves, so this is something I encourage you not to worry about. However, if you plant a terrarium and *all* of the plants start to die off, something clearly isn't right!

As covered earlier in the book, there are a few basic steps you *must* get right to ensure the health and longevity of your terrarium—making sure it gets adequate light, is not overwatered, and high-quality substrate is used will eliminate the vast majority of problems. However, other issues can occur, and the sooner you can spot them, the sooner you can stop them.

Here are a few of the most common problems that terrarium growers experience.

Plants have grown too large

While I don't personally view this as a problem, it's something that's commonly mentioned to me. Thankfully, this is a sign that the plants are happy, not sick!

Remedying this is easy: using a pair of long scissors, prune out some of the larger, older foliage so the plants fit better in the space. (Don't forget to put those cuttings in a propagation box; see page 212!) When pruning terrestrial ferns, select the leaf you wish to prune and track it to the base of the plant and make the cut there. For the vast majority of terrarium plants, it's best to prune slightly above a leaf node—where leaves come out from the stem.

Alternatively, you can just leave the plant to continue growing. I think this can look nice and lead to interesting talking points, but for the long-term health of the plants and the terrarium as a whole, it's best to prune some foliage to ensure light can get into the terrarium.

The leaves are discoloring

This can be a warning sign for many different issues. It's natural for older leaves to discolor and die off, and this is a good thing if you have custodians in the terrarium, as they will feed on any of the decaying leaves. If the substrate you use is of a high quality and full of nutrition, this is something you shouldn't experience much.

Leaf discoloration is not always due to lack of nutrition; a compact substrate, root damage, or an overly wet or dry substrate can also cause this problem. To counteract this, be aware of how often the plants need watering (see page 32) and make sure the substrate is airy and open.

Soil pH can also be a factor for this issue. Some plants prefer acidic conditions, and watering them with regular tap water, which is more alkaline, can cause them to discolor. This isn't a common problem, though, as the majority of terrarium plants aren't overly fussy about soil pH and can handle a little deviation either way. This problem can also be prevented by making sure you water using deionized or distilled water rather than tap water (see page 35).

Pests can also cause leaves to discolor. I've never encountered a pest that enjoys being in the humidity of a terrarium, but if leaves are discoloring and the plants appear sickly, do thoroughly check your plants for signs of pests.

Spotting pests

Telltale signs of pests include but are not limited to:

- Sticky sap that appears on the leaves and sides of the container
- Leaves that are nibbled or eaten
- Leaves that curl inward on themselves
- Clusters of aphids
- The presence of ants—this will often signify an aphid population

RIGHT Pruning a *Ficus pumila* from an overgrown terrarium.

Mold

Terrariums are usually sealed environments with little airflow, and combined with high humidity this can lead to the appearance of mold. Thankfully, I've found that using good-quality substrates, ensuring there is effective air circulation in the container, and planting up healthy plants greatly reduces the risk of mold, and it most often only comes from the presence of wood used for hardscape.

Most types of wood will go through a cycle in a terrarium where they get covered in mold. This can be alarming when you've spent hours making your terrarium and all of a sudden a mold outbreak occurs! But don't worry; mold on wood in a terrarium will nearly always pass on its own. It never spreads to the plants and, other than being unsightly, it is not a real problem. If you want to tackle it, adding a cleanup crew in the form of springtails can help prevent small outbreaks from becoming larger, too.

I have encountered mold issues with certain aquatic substrates, though, which are more of an issue. They stay very wet and a spiderweblike mold can form on the surface, which then spreads to the plants, killing them quickly. To prevent this when using these substrates, I would recommend top-dressing with a layer of bonsai medium like akadama or lava rock, as these seem to keep mold issues at bay.

Slime mold

This is actually a type of fungus, but it's one that will cause no harm to your plants. It forms a fascinating growth on the inside of the container that looks like webbing or veining and spreads outward across the glass, often climbing the sides. The Tokyo subway system appears to model the patterns from this slime mold; in 2010, researchers from Japan and the UK conducted an experiment where they administered nutrients to a slime mold, arranging them to replicate the nodes of the Tokyo subway system. The resulting network bore a striking resemblance to the actual subway system, giving rise to the field now recognized as biologically inspired adaptive network design. Subsequent researchers have conducted studies within the context of their regional rail or road networks. I often see people asking about slime mold in groups online, but I'd advise to leave it be—it isn't problematic in your terrarium and causes no harm, if a little unsightly.

The custodians have died :(

It's a sad day when I look in a terrarium and see that one of my beloved woodlice has died. They do get old, like us, and die of natural causes, but if you notice that your once-booming population of custodians has disappeared, there is likely a problem.

NOT ENOUGH AIR

I often hear people say that the plants within a terrarium produce enough oxygen to keep the custodians alive. I don't know of a way to measure this but I'm sure that isn't the case. Unless you have an exceptionally large terrarium, with exceptionally established plants, colonies of custodians are not going to be happy being in a sealed container. Provide them with tiny air holes, ideally on the sides of the container, but if that's not possible (as in with glass), drill some little air holes in the lid.

TOO LITTLE FOOD

Expecting custodians to feed only on the scraps of dying leaves from a terrarium isn't going to help them thrive. Maybe if they are in huge containers with plants that drop lots of leaves they might have ample amounts of food, but even then, if there is no protein or calcium supply they will become sick and weak. Ensure there is ample amounts of rotting wood as well as leaf litter, and that supplementary feedings are consistent, with small amounts of food given to them at regular intervals throughout the week.

NOT ENOUGH PROTEIN OR CALCIUM

Like us, custodians need nutrients to develop strong exoskeletons, and unfortunately they cannot survive on a vegan diet. A light dusting of calcium powder, which can be bought from any pet store, and a pinch of fish flakes, is enough to keep them healthy and breeding. It's vital that they have these nutrients or their population will quickly diminish.

TOO MUCH FOOD

I'm a feeder, there's no doubt about that—it's definitely my mother's doing! But adding too much food to a terrarium can result in mold issues, and if it's left there this can cause problems for the custodians, especially if it's a small container. Be sure to remove any rotting food before it becomes moldy. It's best to feed little and often, rather than larger amounts less frequently.

OVERLY WET OR DRY CONDITIONS

Different species require different conditions, so it's important to do thorough research before making homes for your custodians. I find it's helpful to create a wetter and a drier area within the terrarium. A wet area can have more moss in it, while the dry area can be filled with leaf litter. If the custodians feel too wet, they will migrate to the dry area: if they feel too dry, they will migrate to the wet area. Simple!

Resources

PLANTS

There are numerous places to find terrarium plants, and visiting local plant shops and garden centers is a good starting point. The baby plants section often contains suitable species, and you can also find larger "mother plants."

Doing some research and finding out the botanical names of plants before you browse is advisable, and a search on Google will often bring up collectors selling their cuttings and plants on Etsy, eBay, and other websites for a fraction of the price. There are many online plant stores that have terrarium sections on their websites, too, or individual collectors who sell offcuts from their collections. The latter is often my favored way of buying rare plants as I learn so much by engaging in conversation with these people!

Facebook groups are another great place to find interesting plants, with users often open to selling or swapping species. In the community section (right), there is a list of terrarium groups that I recommend joining.

Moss

I recommend Moss Clerks. For aquatic species, I go to reputable stockists who purchase from sellers such as Tropica or Dennerle.

SUBSTRATES

For premixed substrates I recommend:

GrowTropicals (UK)
Soil.Ninja (UK and EU)
Josh's Frogs ABG Mix (US)

For bonsai mediums, I recommend buying from bonsai nurseries local to you (buying larger bags is most cost-effective). Here are three in the UK:

Kaizen Bonsai
Bonsai4Me
Herons Bonsai

If you decide to use compost in your terrariums, buy good-quality composts from sellers such as Compost Club.

TOOLS

I don't think you need a big arsenal of tools to build terrariums, but I do recommend buying high-quality ones, if your budget allows.

Tropica and ADA sell great aquascaping tools that work fine in a terrarium setting.

Niwaki sell high-end gardening tools that will last a lifetime.

Try your local Chinese supermarket for a range of different-sized chopsticks!

BOOKS

The books in this selection have been instrumental in my horticulture journey. As such, this book wouldn't have been possible without the time spent reading, watching, and studying these materials over the years and I'd like to express my gratitude to everyone on this list.

Beth Chatto's Woodland Garden, Beth Chatto (Cassell Illustrated, 2002)

Bonsai Inspirations, Volumes 1 and 2, Harry Harrington (B4MePublishing, 2014)

Bonsai Masterclass, Peter Chan (Sterling Pub Co, 1988)

Bonsai Techniques, Volumes 1 and 2, John Yoshio Naka (Dennis Landman Pub, 1998)

Create Your Own Bonsai With Everyday Garden Plants, Peter Chan (Cassell Illustrated, 1992)

Literati Style Penjing: Chinese Bonsai Masterworks, Zhao Qingquan and Thomas S. Elias (BetterLink Press Inc, 2015)

Niwaki, Jake Hobson (Timber Press, 2007)

Not Another Jungle, Tony Le-Britton (DK, 2023)

The Art of Creative Pruning, Jake Hobson (Timber Press, 2011)

The Bonsai Art of Kimura, Katsuhito Onishi (Stone Lantern Pub, 1992)

The Complete Gardener, Monty Don (DK, 2021)

The Modern Japanese Garden, Michiko Rico Nose (Tuttle Pub, 2002)

The New Plant Parent, Darryl Cheng (Abrams Image, 2019)

The Plant Rescuer, Sarah Gerrard-Jones (Bloomsbury, 2022)

COMMUNITY

In order to take your learning further, I think it's important to mix with others who are doing similar things. Online communities are one of the best ways to fast-track your learning.

Facebook groups

- **Anubias, Bucephalandra and Cryptocoryne (ABC) Emerse cultivation**
- **Closed Terrariums**
- **Isopod Breeding and Keeping**
- **Orchids, Epiphytes and Other Plants in Mounted Cultivation**
- **Rare Miniature Terrarium Plants**
- **Springtails**
- **Terrariums/garden in the jars**
- **Terrarium Group**
- **The BioZone: Terrariums, Vivariums, and other Bioactive Setups**
- **Worcester Terrariums**

Social media

Much of the inspiration I have found has come from other creators in the hobby. Here is a list of my favorites, and I've no doubt you'll find loads of inspiration here for yourself.

Another World Terraria
Instagram @another_world_terraria
YouTube @AnotherWorldTerraria

AntScapes
www.antscapes.co.uk
Instagram @ant.scapes
YouTube @AntScapes1
TikTok @AntScapes1

Asu
Instagram @asu_green_jp
YouTube @Asu_green_jp

Bantam Earth
bantam.earth
Instagram @bantam.earth
YouTube @BantamEarth
TikTok @bantam.earth

Benjiplant
Instagram @benji_plant
YouTube @benjiplant
TikTok @benjiplant

Biotope Gallery
Instagram biotope_gallery.official
YouTube @biotopegallery1130

Donny Greens
www.donnygreens.com
Instagram @donnygreens
YouTube @DonnyGreens

Doodle Bird Terrariums
www.doodlebirdterrariums.com
Instagram @doodlebirdterrariums
YouTube doodlebirdterrariums1285
TikTok @doodlebirdterrariums

Dr. Plants
YouTube @TheDrPlants

Ferret Wonderland
Instagram @aquariumferretwonderland
YouTube @ferretwonderland

Homemade Ecosystems
YouTube @HomemadeEcosystems
TikTok @homemade_ecosystems

House Plant Journal
www.houseplantjournal.com
Instagram @houseplantjournal
TikTok @houseplantjournal
YouTube @HousePlantJournal

In Search of Small Things
insearchofsmallthings.com
Instagram @insearchofsmallthings

Kinocorium
Instagram @kinocorium
YouTube @kinocorium

Leafy Street
Instagram @leafystreet_
TikTok @leafystreet
YouTube @leafystreet

Life in Jars?
Instagram @lifeinjars_real
YouTube @LifeinJars

Okopipi
web okopipi.co.uk
Instagram @okopipiuk

Ome Home (Joe Rees)
ome.design
Instagram @ome.home

Pumpkin Beth
www.pumpkinbeth.com
Instagram @pumpkinbeth
YouTube @pumpkinbeth

SerpaDesign (Tanner Serpa)
www.serpadesign.com
Instagram @serpadesign
TikTok @serpadesign
YouTube @SerpaDesign

Terra Jardim
Instagram @terra_jardim

Terrarium Designs (Jordan Polar)
terrariumdesigns.store
Instagram @terrarium.designs
TikTok @terrarium.designs
YouTube @TerrariumDesigns

Terrarium Tribe
terrariumtribe.com
Instagram @terrariumtribe

The Urban Nemophilist (Sushanto Chowdhury)
urbannemophilist.com
Instagram @the_urban_nemophilist
TikTok @urban.nemophilist
YouTube @TheUrbanNemophilist

Walden Plants (Danijela Botaster) walden-plants.hr
Instagram @walden_plants

Index

Page numbers in **bold** refer to captions.

A

Actiniopteris australis 142
Aechmea 190, **190**
aesthetics 56–63
airflow 13, 15, **15**, 154
akadama 24, 31, 79, 113, 184, 216
allotments 6–7, 20, 204–5
amphibians 16–17, 52
Anadenobolus monilicornis 125
Anoectochilus roxburghii 'Red' **183**
Anubias 156
  A. barteri 'Mini Coin' 156, **156**, 159–61
aquariums 13, 70–5, 106–11
aquatic environments 16, 158–61, 216
aquatic plants 13, 16, 154–7, **154**
Armadillidium **17**, 100, 107–11, 120
  A. klughii 'Dubrovnik' 122, **123**
  A. nebula 131–3
  see also woodlice
Asparagus setaceus 18
aspect 45
Asplenium 142, **143**
asymmetry 56, 59, 67
Atlanta Botanical Gardens (ABG) mix 23, 31
auxin 42–3
azalea 96–7

B

backgrounds 59, 103–5
bacteria 28, 127, 199
Barbosella dusenii 187
bark 73, 96
  composted 113
  orchid 30, 31
  see also cork bark
Begonia 171–3, 211
  B. dodsonii 107–11, 172, **172**
  B. polilloensis **172**, 173,
  B. sizemoreae **173**, 173
  B. vankerckhovenii 107–11, **173**, 173
Bilobella braunerae 127
Biophytum 188
  B. sensitivum 18, 131–3, 188, **188**
Bloomsbury Festival 7
Bolbitis 156, **157**
bonsai 58, 62–3
  root pruning 116
  substrates 21, 24–6, 31, 37, 49, 54, 73, 79, 216
  in a terrarium 98, 112–16
  tools 52
Botaster, Danijela 177
bromeliads 190–3
bryophytes 150
Bucephalandra 156–7
  B. caterina 157, **157**, 159–61

C

cacti 12–13, 23, 25, 43
calcium 120
  buildup 35, 150
  supplementary 18, 108, 121, 217
candle holders, lidded **12**, 89, 184–6
carbon dioxide 40, 154
carnivorous plants 154, 198–201
Chamaedorea elegans 'Mini' 194
charcoal
  activated 26, 28, 31, 79
  horticultural 28
  lumpwood 26, 128
Cheng, Darryl 63
Chirita tamiana 63, 194, **195**
chlorophyll 40
Chrysanthemum 43
Cladonia 100
closed systems 10, 20, 37
Cochlidium serrulatum 107–11, 143, **143**
coir 21–4, 26–7, 31, 103, 105
colorful use **56**, 57
communities 218–19
compost 20–4, 27, 31
condensation 38
containers 12, 15, **15**, 49, 52–3, **53**, 88–9
  see also specific containers
cork bark 65, 67, 83, 88, 97, **97**, 99, 104, 108, 110–11, 134–5
costs 61
crickets 134–5, 139
crustaceans 120
Cryptanthus bivittatus 192
Cubaris
  C. murina 18, 122
  C. sp. 'Rubber Ducky' 122, **123**
cultivation 203–17
custodians 13, 16–17, 52, 120–39, 217
cuttings 49, 61, 88, 206–7, 212–13
  Begonia 171
  Fittonia 174, 177–9, **208**
  jewel orchids 182
  leaf 211
  rootless 79
  stem 171, 208, **208**
cuttlefish bone 18, 107–8, 121, 126

D

darkness 41–3
Davis, Miles 56
demijohns 15, 49, 89
Desmoxytes planata 18, 125, **125**, 131–3
detailing roots 98
detritivores 120, 121
Dicksonia antarctica 29
Dicranum
  D. majus 152, **152**
  D. scoparium 38, 41, 150, 152
Dionaea muscipula (Venus flytrap) 199, 200
disease 61
Disney 7, 205
division **207**, 208, 211
Doryopteris cordata 143
Dossinia marmorata 108, 182, **183**
dragon stone 95, **95**, 131–3
drainage 53, 178
driftwood 72, 74, 98, **98**
Drosera (sundews) 199, 200
dry start 154
Dusk Moss Mix 23, 108–10

E

Earth 10
Ebonasea 98
ecosystems 6, 10, **10**, 18, 52, 87, 212
  closed 20, 37
  and minibeasts 17, 121–2, 184
  and mosses 153
  self-sustaining 15, 18
  wetland 16
  and wood 96
Eisenia fetida 127
Elaphoglossum peltatum 107–11, 143, **143**
emergent (marginal) plants 16
energy conservation 43
epiphytes 49, 76, 88, 94, 98–9, 107, 144–5, 166–9, 172–3, 187, 190–1
EpiWeb 99
Equisetum (horsetails) 142
etiolation 41, **42**, 43, 150
exoskeletons 120, 125, 217
expanded clay pebbles 27, 53–4, 66
expanding foam 103–5

F

feeding plants 20, 27
ferns 18, 49, 88, 107–11, 142–5, 205, 208
  aquatic 156

see also specific ferns
fertilizer 27
Ficus 88, 98, 146–9
F. carica 146
F. colombia 18
F. microcarpa 'Ginseng' 113–16, 134–5, 138–9
F. pumila 65, 147, **147**, **148**, 149, **215**
F. punctata 18, 113–16, 131–3, 147, **147**
F. sagittata 149
F. sp. 'Borneo' 146, **146**
F. sp. 'Borneo Small' 91–2, 147, **147**
F. thunbergii 32, 49, **59**, 88, 131–3, 149, 207
F. villosa 149
figurines 86–7, **87**, 91–2
fish 13, 16
fishbowls 64–9, **65**, 130–3
Fittonia 57, 61, 88, 174, **175**, **208**
Colorful *Fittonia* terrarium 176–9
F. albivenis 65–9, **175**, 177–9
flies 134–5, 139
flowering 43
Folsomia candida (springtails) 17–18, **38**, 101, 107–11, 120–1, 127–8, 131–3, 184
foot-candles (fc) 44
foraging 153
foregrounds 59
Frodo stone 94
frogs, dart 17
fronts 58
frosted windows 45
fungi 120, 127, 216

G

geckos 17
glucose 40
Goodyera malipoensis 182, **183**
grain mites 101
Guzmania lingulata var. minor 191

H

hardscape materials 65, 93–101
see also specific materials
Hedera helix 'Minima' 194, **195**
Hemionitis arifolia 143
hermaphrodites 126
horticulture 6–7, 204
human impact 10
humidity 12–13, 15, 38, 75, 108, 121–2, 154, 216
Hydrocotyle 157
hydrophilic materials 99
hydrophobic substrates 21, 22, 26
Hygrolon 99
Hymenopus coronatus **135**, 138–9
Hypnum cupressiforme 152
Hypoestes 194

I

impact, creation 61
interconnectedness 10
isopods 17, **17**, 93, 100, 107–11, 120–2, **123**, 131–3

K

Kita Mountain rock 94
kyodama 25

L

Latimer, David 15, 18, 196
lava rock 24–5, 31, 49, 74, 79, 94, **94**, 113, 159–61, 216
layering 87
leaf cuttings 211
leaf litter 101, **101**, 108, 125, 133
leaf mold 20, 28, 184
leaves
dead **38**, 63
discolored 215
LEGO® 131, 133
Lemmaphyllum microphyllum 145, **145**
Leucobryum glaucum 38, 65–9, 81–3, 91–2, 113–16, 152–3, **152**, 211, **211**
preserved 100
lichen 100
lids 53, 69, 83, 107, 111, 138–9
light 15, 40–7
and carnivorous plants 199–200
Low-light terrarium 184–6
measurement 44, **44**
and mosses 150
and propagation 212
requirements 41–3
sources 45–6, **46**
light meters 44, **44**
literati/bunjin style 62

M

Macodes petola 182, 184–6
malachite 100
manure 20
Marcgraf, George 169
Marcgravia 88, 169–70, **170**
M. 'Mini Limon' 107–11
Marvel 7, 205
mental health 6, 205
mesh 54, 66, 132, 138–9
micro terrariums 88
Microgramma vacciniifolia 144, **144**
midgrounds 59
millipedes 17–18, 93, 100, 121, 125, **125**, 131–3
Mnium hornum **32**, 153
moler clay 24–5, 31, 54, 113
Monet, Claude 204
mossariums 15
mosses 12, 15, 57, **59**, 80, 89, 107–8, 115, 150–3, 218
aquatic **12**, 15, 38, 151, 151, 211
from the wild 153
and light levels 41
preserved 100
propagation 211, **211**
terrestrial 152–3, **152–3**
watering 38
see also specific mosses
mold 43, 96–7, 120, 127, 216
mycelium 22

N

naturalistic terrariums **56**, 57
Neoregelia 192, **192**
Nepenthes (pitcher plants) 199, **199**, 200, 201
Nephrolepis **42**, 45, 65–9, 131–3
N. cordifolia 'Duffii' 18, 91–2, 144, **144**, **207**
N. exaltata 71–5, 144
nutrient cycle 10, 17, 18, 120

O

orb containers 90–2
orchid bark 30, 31
orchid mantis 135, 138–9
orchids 187
jewel 41, 45, 89, 108, 182–6, **183**, 187
micro 88
Phalaenopsis 45, 187
Oxychilus alliarius (garlic snail) 18, 126, **126**
oxygen 17–18, 40, 121, 217

P

paludariums 16, 52, 76, 168
parthenogenesis 127
peat 22, 200, 206
Pellaea rotundifolia 145
Pellionia repens 107–11, 195
penjing 63
Peperomia 49, 166–8, 211
P. antoniana 166, **166**
P. emarginella 81–3, 167, **167**
P. hispidula 167, **167**
P. 'Pepperspot' 49
P. prostrata 71–9, 91–2, 168, **168**, 211
P. verticillata 61, 168
Perionyx excavatus 127
perlite 22, 200
pests 215
Phalaenopsis 45, 187
Phidippus regius 134–6, **135**
Philodendron 41, **180–1**, 180–1, 205
P. pteropus 'Mini' 181, **181**
P. sp. 'Mini Santiago' 180, **180**
P. verrucosum 'Dwarf' 181, **181**
P. verrucosum 'Mini' 181
photoperiodism 42
photosynthesis 40–2, 61, 150
Phyrrobryum dozyanum **153**, 153

physical exercise 205
phytochrome 42
Pilea 88, 164–5
P. cadierei 134–5, 138–9, 164, **165**
P. glauca 65–9, 164, **165**
Pinguicula 201, **201**
Plagiomnium affine 151
Plant, Benji 151
plants 141–201, 218
choice of 88
death 63, 214
decaying matter **38**, 120–2, 127
growth 61, 88, 214, **215**
rare 41, 52, 61, 88, 122, 144, 156, 165, 168, 170, 172–3, 180, 187, 206–7, 218
see also specific plants
pollution 10, 87
Porcello laevis (dairy cows) 122, **123**
Praying mantis terrarium **135**, 138–9
propagation 49, 61, 88, 171, 206–13, **207**
see also cuttings
propagation boxes 207, 212–13
protein 18, 120, 122, 125–6, 217
pruning 61, 75, 116, 214, **215**
Psilotum 142
pteridophytes 142
Pteris 145
pumice 24, 25
Pyrrhobryum dozyanum **32**
Pyrrosia 88
P. nummularifolia 49, 145

R

reindeer moss (lichen) 100
reptiles 17, 52
rheophytes 156
rhododendron 96
RHS Chelsea Flower Show 7, 18, 205
ripariums 16
rocks 94–5, **94–5**
Rule of Thirds 60, **60**

S

sakadama 24
sand 30
Saxifraga stolonifera 'Mini' **12**, 195, **195**
Schwartz, Matthew 107
seals 15, 121
seasons 45
seedpods/cones 100
seeds 208
Seiryu stone 95, **95**, 134–5
Selaginella 41, 45, 89, 162–3
S. erythropus 163
S. uncinata 107–11, 162–3, **163**, 184–6
S. willdenowii 162–3, **163**
self-care 204–5
shale 94
shrimps 13, 16
silicone 103–5
Singapore Botanic Gardens 6
Sinningia muscicola 196, **196**
slate 94
slime mold 216
slugs 101
snails 16–18, 101, 121, 126, **126**
soil 20–31, **21**
aerobic 120
anaerobic 21, 32, 34
garden 22
moisture levels 32
pH 95, 215
see also watering
Solanum sp. 'Ecuador' 107–11, 196
Specklinia dressleri 88, 187
sphagnum moss 23–4, 31, 54, 88, 99, 103, 105, 108–9, 111, 113, 138–9, 142, 171, 180, 182, 184, 186, 200, 212–13
spice jars 80–3, 89
spider wood (azalea root) 96–7, **97**
spiders 17, 134–6, **135**
Spinacia oleracea 43
Spirits bottle terrarium 76–9
spores 142, 162
springtails *see Folsomia candida*
sterilization 22, 93, 101
Streptocarpus ionanthus (African violet) 196, 211
substrates 20–31, **21**, 49, 54, 218
aerobic 120
anaerobic 21, 32, 34
aquatic 159–61, 216
to avoid 22
and building terrariums 66, 72–3, 79, 82, 92, 113–14, 116, 132, 135, 139, 159–61, 178–9, 186
and carnivorous plants 200
compaction 215
components 24–30
DIY 31
high-quality 20–1, 23
hydrophobic 21–2, 26
pH 24–5, 95, 121, 215
premixed 22, 23
and propagation boxes 212–13
sterilization 22
sustainable 24
water-retentive 20–1
well-draining 20–1
see also bonsai substrates
succulents 12, 13, 24, 25, 43
symbiosis 17
symmetry 56, 59, 67

T

Taxiphyllum 107–11
T. barbieri 12, 38, 71–5, 151
T. 'Spiky Moss' 151, **151**
terrariums
Aquarium 70–5
Aquatic plant 158–61
bioactive 12, 17, **17**, 101, 107–11, 130–3
Bonsai 98, 112–16
Colorful *Fittonia* 176–9
cultivation 203–17
definition 12
Fallen branch 106–11
famous 18–19
getting started 51–83
Globe 90–2
Jumping spider 134–6, **135**
and light 40–7
Low-light 184–6
old 15, 18, 48–9, **49**
ongoing care 75, 80, 186, 214–17
open 12, 13
personalization 86–92, **87**
plants 141–201
Praying mantis **135**, 138–9
principles of 9–49
seals 15, 121
Simple fishbowl 64–9, **65**
Spice jar 80–3
Spirits bottle 76–9
types of 13–17
see also custodians; plants; substrates; watering
Thuidium tamariscinum 150, 153
Tic Tac boxes 88
Tillandsia stricta 138–9, 193, **193**
tools 52, 55, 218
tortoises 17
Tradescantia 15, 196
tree fern fiber 29, 31, 97, 99, 110, 184
Trichorhina tomentosa 122

U

upcycling 52, 59
Utricularia 199
U. graminifolia 154, **154**

V

Vesicularia
V. dubyana 38
V. montagnei 107–11, 151, 159–61
vivariums 16, 17, **46**, 52, 98–9, 190
Vrydagzynea tristriata 182

W

wabi-sabi 63
Walstad method 13
water cycle 10, 12
water propagation 208
watering 32–9, 40, 186, 215
bromeliads 190
and building terrariums 66, 74, 82, 113, 132–3, 160, 178
carnivorous plants 200
how to 34, 37, **37**
misting **32**, 37–8
open terrariums 13
overwatering 32–4, 113
types of water 35
water pH 150
white panther stone 94
wood 96–8, **96–8**, 216
decaying 125
grape 83, 96
Malaysian driftwood/ blackwood 98
Manzanita 96
mopani 98, **98**
petrified 96, **96**, 115
rotting 100, 121, 131
woodlice **17**, 18, 100, 107–11, 120–2, **123**, 131–3, 217
worm castings 23, 29, 31, 113, 127, 184
wormeries 29
worms 127

Z

zeolites 30

Author's acknowledgments

It wouldn't be right to place anyone ahead of my mother on the acknowledgments page of this book. She is the person who has stuck by me through my darkest days; from when Dad passed away at the turn of the millennium, to my brother Stephen's death just a few years later, and through the turbulent years that followed where I was unfathomably difficult. We've been through an incredible amount and throughout all of this, she has been an unwavering pillar of support. Without her, none of this would be possible and I can't put into words how much I owe her.

Next is my sister, Tanya, who, from ordering me taxis when I was lost in the middle of nowhere and checking my grammar on a ridiculously large number of emails, letters, CVs, and job applications has been someone I've always looked up to. She is always offering her insight and telling me what she believes is right, even if it's not what I want to hear.

To my father, who will never read these words, and who I never properly got to know. I've been told about your fascinating life and not hearing those stories from you is one of my biggest regrets. Even though our time together was short, I must say a special thank you to you.

To Grace, my girlfriend, who inspires me every day. You always provide grounding advice and feedback, and you make me feel good about myself, especially during periods of doubt. Your unwavering belief in me makes all the difference. Thank you for being both a pillar of strength and a source of inspiration. I'm grateful for all you bring to my life.

To Chris Young, Ruth O'Rourke, Lucy Philpott, Barbara Zuniga, Helena Caldon, and photographer Jason Ingram who have made writing this book the most enjoyable experience.

My agents, Crystal and Jason, for believing in me and being by my side on this wonderful journey. My agents, Crystal and Jason, for believing in me and being by my side on this wonderful journey.

To the RHS Chelsea Flower Show team of Sarah Gerrard-Jones, Mark Lawlor, Ian Morrison, Amy Loosely, Jack Binns, Levent Latif-Maeer, and Jamie Gregson. We won a gold medal!

I am deeply grateful to Andy Edwards, my music teacher at Kidderminster College, who I met in 2011. Meeting him marked a turning point in my life. His incredible drive and dedication to music deeply inspired me during my brief time there, and his profound influence transformed my perspective and approach to my own passions.

Thank you to Tony Le-Britton, author of *Not Another Jungle* and owner of Not Another Jungle plant store, for recommending me to the DK team.

Many thanks to my friends and those who have inspired me: Dave Powell, Akib Salis, Chris Minton, Stuart Godwin and the rest of my friends at Royal Mail. Alessandro Vitale, Sarah Gerrard-Jones, Ian Morrison, Mark Lawlor, Ofran Al-Mossawi, Don DeLillo, Haydn Rogers, Luigi Abate, Pasquale Abate, Luke Orban, Casper, Dmitry Konev, Elizabeth Humphries, Derek Sarno, Richard Kupfer (we miss you so much) and of course, my kind and supportive friends from Worcester.

I would like to add a special mention to Danijela Bota, Delia La Cruz, Sebastian Irie and Aurélie Morgen, who are the admins of my Facebook group. Without their work, the group wouldn't be where it is today. Thank you so much.

A special mention to the people and businesses that have supported me over this whole journey: Jacob and the team at GrowTropicals, Daley from The Jungle Club Plant Shop, Joe from Scaped Nature, Riverwood Aquatics, Biorb, Jon Carter from Neotropic Treasures, Adam Webb; chief bug guy at Micro Exotics, Immuto Films, Mitch from Unseen Universe, and the amazing team at Soil.Ninja.

Publisher's acknowledgments

DK would like to thank Adam Brackenbury for repro work on the images, Kathryn Glendenning for proofreading, and Lisa Footitt for creating the index.

About the Author

Ben Newell started creating terrariums in 2016 after discovering the physical and mental benefits of plants and gardening in his mid-20s. Documenting his terrarium journey on social media, Ben has found an audience of over 2.8 million plant enthusiasts across the world who delight in his curiosity and care for nature. In 2022, Ben received a RHS Chelsea Gold Medal in recognition of his work and was short-listed for Garden Media Guild Social Media Influencer of the Year. He has collaborated with brands, including Disney and Marvel, and been featured in The Spruce, BuzzFeed, and LADbible, and made appearances at BBC Hereford & Worcester, Bloomsbury Festival, Great Ormond Street Hospital, and on Michael Perry's and Ellen Mary Webster's The Plant Based Podcast.

Disclaimer

The plants in this book have been referred to according to internationally recognized scientific (or "Latin") naming conventions. DK acknowledges that these conventions do not necessarily represent the true origin and cultural significance of the plants in question, which may be known by a number of names, including those applied to them by Indigenous groups in the locations in which they first grew.

Editorial Manager Ruth O'Rourke
Project Editor Lucy Philpott
Senior Designer Barbara Zuniga
Senior Production Editor Tony Phipps
Production Controller Kariss Ainsworth
DTP and Design Coordinator Heather Blagden
Jacket Coordinator Emily Cannings
Art Director Maxine Pedliham
Publishing Director Katie Cowan

Project Editor Helena Caldon
Design Evi-O.Studio
Photography Jason Ingram
Consultant Gardening Publisher Chris Young

First American Edition, 2024
Published in the United States by DK Publishing, a division of Penguin Random House LLC
1745 Broadway, 20th Floor, New York, NY 10019

24 25 26 27 28 10 9 8 7 6 5 4 3 2 1
001–340511–May/2024

Published in Great Britain by Dorling Kindersley Limited

A catalog record for this book is available from the Library of Congress.
ISBN: 978-0-7440-9907-2

Printed and bound in Dubai
www.dk.com

This book was made with Forest Stewardship Council™ certified paper – one small step in DK's commitment to a sustainable future. Learn more at **www.dk.com/uk/information/sustainability**